AF361920

Plasma-Catalysis for Environmental and Energy-Related Applications

Plasma-Catalysis for Environmental and Energy-Related Applications

Editors

Monica Magureanu
Corina Bradu

MDPI • Basel • Beijing • Wuhan • Barcelona • Belgrade • Manchester • Tokyo • Cluj • Tianjin

Editors

Monica Magureanu
Department for Plasma Physics
and Nuclear Fusion
National Institute for Lasers,
Plasma and Radiation Physics
Măgurele-Bucharest
Romania

Corina Bradu
Department of Systems Ecology
and Sustainability
University of Bucharest
Bucharest
Romania

Editorial Office
MDPI
St. Alban-Anlage 66
4052 Basel, Switzerland

This is a reprint of articles from the Special Issue published online in the open access journal *Catalysts* (ISSN 2073-4344) (available at: www.mdpi.com/journal/catalysts/special_issues/plasma_catal).

For citation purposes, cite each article independently as indicated on the article page online and as indicated below:

LastName, A.A.; LastName, B.B.; LastName, C.C. Article Title. *Journal Name* **Year**, *Volume Number*, Page Range.

ISBN 978-3-0365-2783-3 (Hbk)
ISBN 978-3-0365-2782-6 (PDF)

Contents

About the Editors

Monica Magureanu

Dr. Monica Magureanu is a plasma physicist working in the National Institute for Lasers, Plasma and Radiation Physics in Bucharest, Romania. Her research activities are mainly related to experimental investigations of non-thermal plasma for various applications, such as water treatment (i.e., the degradation of organic contaminants in water), plasma agriculture (seed treatment), plasma catalysis for the oxidation of volatile organic compounds in air, and plasma treatments of materials.

Corina Bradu

Dr. Corina Bradu is an environmental chemist and holds a position as an Associate Professor at University of Bucharest, Department of Systems Ecology and Sustainability. Her research focuses on water and wastewater treatment using various techniques, such as advanced oxidation processes for the degradation of persistent organic pollutants, selective catalytic reduction of nitrate and nitrites ions and adsorptive processes for the removal of heavy metals.

Editorial

Catalysts: Special Issue on Plasma-Catalysis for Environmental and Energy-Related Applications

Monica Magureanu [1],* and Corina Bradu [2],*

1 Department of Plasma Physics and Nuclear Fusion, National Institute for Lasers, Plasma and Radiation Physics, Atomistilor Street 409, 077125 Magurele, Romania
2 PROTMED Research Centre, Department of Systems Ecology and Sustainability, University of Bucharest, Spl. Independentei 91-95, 050095 Bucharest, Romania
* Correspondence: monimag@gmail.com (M.M.); corina.bradu@g.unibuc.ro (C.B.); Tel.: +40-766-457-701 (C.B.)

Citation: Magureanu, M.; Bradu, C. *Catalysts*: Special Issue on Plasma-Catalysis for Environmental and Energy-Related Applications. *Catalysts* **2021**, *11*, 1439. https://doi.org/10.3390/catal11121439

Received: 22 November 2021
Accepted: 24 November 2021
Published: 26 November 2021

Publisher's Note: MDPI stays neutral with regard to jurisdictional claims in published maps and institutional affiliations.

Plasma-catalysis has been a topic of research for many years due to its potential for applications in a wide range of chemical, environmental, and energy-related processes. Non-thermal plasma offers an unconventional way to initiate chemical reactions in gas and in liquid due to the energetic electrons generated in the plasma, however, it suffers from low selectivity. The coupling of plasma with catalysis can steer the reactions in the desired direction, thus providing improved selectivity towards the target products and reducing unwanted ones.

This special issue demonstrates the interest in plasma catalysis as solution to environmental problems caused by greenhouses gases CO_2 and CH_4 [1,2] that can be converted to value-added products and fuels, gas pollution with stable polycyclic aromatic hydrocarbons [3] and volatile organic compounds [4], as well as water pollution [5]. The special issue includes a review article and four articles devoted to particular applications.

The review by Gao et al. [1] presents recent developments of plasma catalytic dry reforming of methane (DRM), focusing on dielectric barrier discharge (DBD) reactors in combination with nickel-based catalysts, currently considered as very promising catalysts in DRM reactions due to good catalytic activity, high availability and low cost. Results obtained using monometallic Ni catalysts on various supports (SiO_2, Al_2O_3, ZrO_2 and activated carbon), as well as using Ni-based catalysts doped with other metals or compounds are summarized. The authors discuss the role of the support that can lead to different performance in the plasma catalyzed DRM reaction and present advantages and drawbacks of each support under plasma conditions, related to coke formation, metal sintering in connection with the size of Ni grains, dispersion of the Ni particles on the support, specific surface area in connection with reactant adsorption etc. Doping with other metals or metal oxides (alkali and alkaline earth metals—K and Mg, transition metals—Co and Mn, rare earth metals—La and Ce and their oxides) may prove beneficial, producing various effects such as enhancing resistance to carbon formation and even removing carbonaceous deposits, improving the Ni dispersion and inhibiting aggregation of Ni particles, strengthening the metal-support interaction, enhancing reactant adsorption, changing discharge characteristics as to increase the plasma area etc., ultimately leading to better plasma-catalyst synergy, higher CO_2 and CH_4 conversion and improved products distribution. Another important part of the paper is dedicated to plasma reactor design, including plasma-catalyst coupling, and process parameter optimization. The key parameters impacting on system performance and efficiency, i.e., discharge input power, feed flow rate, specific input energy, and feed ratio, are discussed in detail. The in-depth understanding of plasma-catalyst interaction and of the correlation between efficiency and reactor design and process parameters still remains one of the challenges to further improve this application.

Packing various materials inside the plasma reactor can influence the discharge in many different ways, obviously depending on the nature of the material, its dielectric constant, surface roughness, thermal and electrical properties, but also on their size and shape. Uytdenhouwen et al. [2] explored the potential of core-shell structured spheres to allow tuning of the packing properties for a specific application, by designing an optimal combination of core and shell. They investigated different core-shell combinations, using dense spheres of SiO_2, Al_2O_3, and $BaTiO_3$ as core and spray-coated powders of these materials as shell, and tested the materials for CO_2 conversion in comparison with pure, dense spheres. The authors observed a clear effect of the size of the spheres on the conversion and energy efficiency, and the data showed that this effect is material-dependent. It was also found that the choice of core and shell materials is extremely important to the packed-bed reactor performance and further optimization can be carried out to some extent by adjusting shell thickness. Although it was not the purpose of this work to achieve high activity for CO_2 splitting, as the tested materials are not catalytically active, nor to identify the optimum core-shell structure leading to significant performance improvement, the paper clearly demonstrates great potential in preparing packing materials appropriate for a certain application. Doping the shell with suitable catalysts, as suggested by the authors, represents a very interesting opportunity in this direction due to the macroporous nature of the material that may allow plasma to penetrate inside the pores and thus enhance plasma-catalyst interaction.

Various packing materials in a DBD reactor have also been investigated in the paper by Cimerman et al. [3] who reported the removal of tars, in particular naphthalene, by plasma catalysis in such a system. As previously mentioned, the packing can influence the discharge properties through several mechanisms, and this is ultimately visible in the variation of discharge power as a function of applied voltage. The authors focused on the effect of packing material on naphthalene removal and on the energy efficiency, considering the type of material, the specific surface area, the shape and size of the pellets. The pollutant removal was clearly enhanced when using the packed-bed reactor as compared to the empty DBD reactor, following the sequence TiO_2 (88%) > $Pt/\gamma Al_2O_3$ (78%) > ZrO_2 (72%) > γAl_2O_3 (66%) > glass beads (64%) > $BaTiO_3$ (51%) > plasma only (41%). It was suggested that a key factor for obtaining high removal efficiency is the presence of surface discharges that propagate along the packing material and thus activate a larger area of the catalyst and favor chemical reactions, while high dielectric constant materials, such as $BaTiO_3$, concentrate the discharge close to the contact points of the pellets, resulting in weaker chemical effects. Increasing the specific surface area of the materials favored naphthalene removal, this behavior being similar to thermal catalysis, where textural properties play a major role. However, contrary to conventional catalysis, the shape and size of the packing material were found to be important in plasma catalysis, as they determine the size of voids between the pellets, and thus discharge properties, gas residence time and plasma volume. The authors identified carbon monoxide, carbon dioxide, water and formic acid as main products in the gas phase of naphthalene decomposition, as well as several other complex gaseous and solid by-products, indicating incomplete oxidation of the target compound.

Although recently significant research efforts have been devoted to explaining the mechanisms of plasma-catalyst interaction, the present understanding still leaves open questions related to the complex phenomena involved in plasma catalysis. In order to fill some of these gaps, the article by Vega-González et al. [4] focuses on the reaction pathways occurring during the decomposition of a volatile organic compound (acetaldehyde) by DBD fluidized bed with Ag/TiO_2 catalyst deposited on 150-μm-diameter SiO_2 pellets. The authors obtained enhanced removal of the VOC and considerably higher selectivity to CO_X ($CO + CO_2$) in the plasma-catalytic system as compared to plasma alone. For the highest specific input energy used (1150 J/L), acetaldehyde was almost completely removed (98%) and the selectivity to CO_X reached 60%. Other organic by-products, such as methanol, acetic acid, methyl formate, methyl acetate, 1,2-ethanediol mono- and di-formate, were identified in the gas phase in both reactors as a result of the target compound decomposition, while acetone and nitromethane were only detected in the plasma-catalytic process. In-situ diffuse-reflectance infrared Fourier transform spectroscopy (DRIFTS) was used to monitor the catalyst surface during acetaldehyde adsorption and subsequent plasma degradation. It was found that at the end of the adsorption step the $Ag/TiO_2/SiO_2$ was mainly covered with acetaldehyde, but other species such as acetate, acetone and formaldehyde were also formed. The measurements carried out on the catalyst exposed to plasma reveal that acetaldehyde degradation proceeds with the formation of different surface compounds such as acetate, methoxy, ethoxy, formate and carbonate species, as well as formaldehyde and formic and acetic acid. In addition to the formation of these species on the catalyst surface, some of them can also be formed directly in the gas phase and subsequently adsorb on the catalyst. Detection of the carbonaceous intermediates provided more detailed information on acetaldehyde degradation in the plasma catalytic system by the interaction of adsorbed molecules with plasma-generated species and/or by-products of acetaldehyde decomposition in the gas phase. Thus, the main degradation pathways could be identified.

The degradation of aqueous pollutants is also a challenging issue that may be tackled by non-thermal plasma. Although the removal of target contaminants appears feasible, mineralization is often very slow and energy-consuming, as numerous organic by-products are formed, similarly to the case of gaseous pollutants. Again, the addition of catalysis is believed to provide a solution to this problem and various materials have been tested for this purpose in combination with plasma. Korichi et al. [5] addressed the issue of mineralization for a solution containing paracetamol treated by a DBD above liquid combined with iron catalysts supported on glass fiber and immersed in the liquid. Complete removal of the contaminant was achieved after 60 min treatment by plasma alone at very low input power (0.3 W), but the carbon remained in organic form and no mineralization was obtained. Introducing the Fe^{3+} catalyst proved beneficial to the degradation process: the rate constant increased by more than 40% as compared to the treatment by plasma alone, indicating faster paracetamol removal, and the energy yield was almost two times higher, clearly showing a considerable improvement in the process efficiency. Moreover, the mineralization degree increased to 30% after 15 min and to 54% after 60 min treatment by plasma catalysis, and a decreasing trend with treatment time was observed for most of the detected organic by-products. The role of the catalyst is thus decisive in improving the degradation process and was attributed to the generation of additional hydroxyl radicals by the Fenton-like reaction, i.e., by decomposition of the plasma-produced hydrogen peroxide in the presence of Fe^{3+}. Hydroxyl radicals are highly reactive and generally believed to be the driver of the degradation of organic compounds in water in all advanced oxidation processes. Therefore, the plasma-Fenton combination, using the continuous supply of H_2O_2 generated in-situ by the plasma, shows great potential for water cleaning.

We can conclude that plasma catalysis is very active research area, various applications are envisaged and promising results are reported worldwide. Finally, we wish to thank all the authors for their valuable contributions, without which this special issue would not have been possible.

Conflicts of Interest: The authors declare no conflict of interest.

References

1. Gao, X.; Lin, Z.; Li, T.; Huang, L.; Zhang, J.; Askari, S.; Dewangan, N.; Jangam, A.; Kawi, S. Recent developments in dielectric barrier discharge plasma-assisted catalytic dry reforming of methane over Ni-based catalysts. *Catalysts* **2021**, *11*, 455. [CrossRef]
2. Uytdenhouwen, Y.; Meynen, V.; Cool, P.; Bogaerts, A. The potential use of core-shell structured spheres in a packed-bed DBD plasma reactor for CO_2 conversion. *Catalysts* **2020**, *10*, 530. [CrossRef]
3. Cimerman, R.; Cíbiková, M.; Satrapinskyy, L.; Hensel, K. The effect of packing material properties on tars removal by plasma catalysis. *Catalysts* **2020**, *10*, 1476. [CrossRef]
4. Vega-González, X.; Duten, S. Sauce, Plasma-catalysis for volatile organic compounds decomposition: Complexity of the reaction pathways during acetaldehyde removal. *Catalysts* **2020**, *10*, 1146. [CrossRef]
5. Korichi, N.; Aubry, O.; Rabat, H.; Cagnon, B.; Hong, D. Paracetamol degradation by catalyst enhanced non-thermal plasma process for a drastic increase in the mineralization rate. *Catalysts* **2020**, *10*, 959. [CrossRef]

 *catalysts* 

Review

Recent Developments in Dielectric Barrier Discharge Plasma-Assisted Catalytic Dry Reforming of Methane over Ni-Based Catalysts

Xingyuan Gao [1,2,†], Ziting Lin [1,†], Tingting Li [1], Liuting Huang [1], Jinmiao Zhang [1], Saeed Askari [3], Nikita Dewangan [3], Ashok Jangam [3] and Sibudjing Kawi [3,*]

1   Department of Chemistry, Guangdong University of Education, Guangzhou 510303, China; gaoxingyuan@gdei.edu.cn (X.G.); lziting@gdei.edu.cn (Z.L.); ltingting15@gdei.edu.cn (T.L.); huangliuting@gdei.edu.cn (L.H.); zhangjinmiao@gdei.edu.cn (J.Z.)
2   Engineering Technology Development Center of Advanced Materials & Energy Saving, Emission Reduction in Guangdong Colleges and Universities, Guangzhou 510303, China
3   Department of Chemical and Biomolecular Engineering, National University of Singapore, Singapore 117585, Singapore; e0554257@u.nus.edu (S.A.); dnikita@u.nus.edu (N.D.); chejang@nus.edu.sg (A.J.)
*   Correspondence: chekawis@nus.edu.sg; Tel.: +65-6516-6312
†   The authors contribute equally to this review.

Abstract: The greenhouse effect is leading to global warming and destruction of the ecological environment. The conversion of carbon dioxide and methane greenhouse gases into valuable substances has attracted scientists' attentions. Dry reforming of methane (DRM) alleviates environmental problems and converts CO_2 and CH_4 into valuable chemical substances; however, due to the high energy input to break the strong chemical bonds in CO_2 and CH_4, non-thermal plasma (NTP) catalyzed DRM has been promising in activating CO_2 at ambient conditions, thus greatly lowering the energy input; moreover, the synergistic effect of the catalyst and plasma improves the reaction efficiency. In this review, the recent developments of catalytic DRM in a dielectric barrier discharge (DBD) plasma reactor on Ni-based catalysts are summarized, including the concept, characteristics, generation, and types of NTP used for catalytic DRM and corresponding mechanisms, the synergy and performance of Ni-based catalysts with DBD plasma, the design of DBD reactor and process parameter optimization, and finally current challenges and future prospects are provided.

Keywords: plasma; dry reforming of methane (DRM); dielectric barrier discharge (DBD); Ni-based catalyst

Citation: Gao, X.; Lin, Z.; Li, T.; Huang, L.; Zhang, J.; Askari, S.; Dewangan, N.; Jangam, A.; Kawi, S. Recent Developments in Dielectric Barrier Discharge Plasma-Assisted Catalytic Dry Reforming of Methane over Ni-Based Catalysts. *Catalysts* **2021**, *11*, 455. https://doi.org/10.3390/catal11040455

Academic Editors: Monica Magureanu and Corina Bradu

Received: 26 February 2021
Accepted: 29 March 2021
Published: 1 April 2021

Publisher's Note: MDPI stays neutral with regard to jurisdictional claims in published maps and institutional affiliations.

1. Introduction

With industrial development and social progress, a large amount of fossil fuels have been burned, energy consumption in various places has increased sharply, carbon dioxide emissions have increased greatly, and the greenhouse effect has become a serious issue, so reducing the negative impact of climate change and slowing down global energy consumption are of vital importance. At the same time, greenhouse gases (CH_4 and CO_2) are also deemed as the raw feed for production of value-added chemicals [1,2]. Therefore, dry reforming of methane (DRM) has been put into industrial production and has attracted widespread attention because of its dual benefits of environmental protection and resource utilization [3–7]. The product is generally synthesis gas or other valuable chemical substances, such as hydrogen and carbon monoxide (syngas components), ammonia, methanol, acetic acid, synthetic gasoline, etc. The main reaction equation is:

$$CH_4 + CO_2 = 2CO + 2H_2, \Delta H_{298K} = 247\ kJ/mol$$

because of the high stability of methane and carbon dioxide molecules. Studies have shown that DRM has the following advantages: (1) Wide sources of raw materials are utilized to turn waste into treasure, thereby reducing atmospheric pollution; (2) compared to wet reforming and partial oxidation reforming of methane, DRM can save nearly half of methane; (3) the ratio of H_2 to CO is close to 1, appropriate for oxo reaction (hydroformylation reaction in the presence of Co or Rh where olefins reacts with CO and H_2 to produce aldehydes) and FTS reaction (also known as Fischer-Tropsch Synthesis, a process where CO and H_2 react to form olefins and other valuable products); (4) the reaction has a large reaction heat, which can be used as energy storage and medium transmission, such as solar energy storage [1,8–17].

However, in recent years, the traditional DRM method has gradually shown some limitations, such as high energy consumption and catalyst deactivation at high operation temperatures [9,18]. In this case, the emergence of plasma technology overcomes the high energy input to activate methane and carbon dioxide molecules. The high-energy electrons produced by plasma can initiate chemical reactions at room temperature, providing a new way for reforming reactions. Traditional methane conversion underwent a two-step catalytic process, usually conducted under intensive conditions. In contrast, non-equilibrium plasma-assisted DRM where electrons possess a higher energy owns a much lower reaction temperature than that of the traditional reforming process. In other words, the non-equilibrium plasma has the potential to generate active species in situ at low temperatures to accelerate the reaction process [8,19].

Non-thermal plasma technology (NTP) has attracted more and more attention due to its simple equipment, easy operation, high conversion rate; moreover, NTP saves the energy and materials and benefits the environmental protection. Current studies have shown that a variety of NTP structures have been tested for DRM reactions, such as gliding arc discharge, dielectric barrier discharge (DBD), microwave discharge, corona discharge,, etc. Among them, DBD has drawn special attentions because of its high density of electrons and production of highly active species. Scientists have conducted a lot of research on related areas [8,10,20–22].

Despite many advantages of DBD in DRM, the reaction has a lower activity and serious coke formation without a catalyst. Integrating NTP with the catalyst, a synergistic effect is observed to increase the conversions of reactants and the selectivity or yield of the targeted product due to the specific excitation formed in the plasma with sufficient energy. In detail, the electric field of plasma is enhanced due to charge accumulation and a polarization effect caused by the roughness of the catalysts. Additionally, a higher interaction between the catalyst and active species causes a higher conductivity, which improves the magnitude of the electric field. Meanwhile, the catalyst facilitates the adsorption and prolongs the contact time of reactants, leading to a higher activation and conversion. On the other hand, the highly reactive species generated by plasma cause the structure change and surface faceting of the catalyst, promoting the charge deposition and hotspot formation, which activates the metals and reactants. The combination of plasma and catalysts can prolong the lifetime of active species and lower the activation barrier [8]. Among the catalysts for plasma-catalyzed DRM, noble metals show good catalytic performances. As seen from previous studies, precious metals show strong resistance to coke formation and exhibit high catalytic activity in thermal and plasma DRM. The precious metals commonly used in plasma DRM include Pt, Pd, Rh, and Ru [8]. For example, Pt nanoparticles were impregnated in the UiO-67 MOF structure, accelerating the reactant dissociation and enhancing H_2 yield. Moreover, due to the dehydrogenation of hydrocarbons on Pt nanoparticles, the selectivity towards light hydrocarbons was reduced by 30%. Assisted by plasma, the surface reactions were intensified, increasing the energy efficiency by 11%. Furthermore, excellent stability was shown by the constant conversions during four cycles of plasma on-off [23]. However, due to their scarcity, high price, and easy sintering at high temperatures, they are not suitable for large-scale industrial applications [24,25]. Relatively speaking, transition metals such as Co, Mn, Fe, Cu, and Ni are also used for plasmonic catalytic DRM reactions. Zeng et al. compared the performance of different γ-Al_2O_3 supported metal catalysts M/γ-Al_2O_3 (M = Ni, Co, Cu, and Mn) in the plasma assisted DRM. They found out that the combination of plasma with Ni/γ-Al$_2$O$_3$ and Mn/γ-Al$_2$O$_3$ catalysts significantly increased CH$_4$ conversion [26,27]. Other studies have shown that among various transition metals (Ni, Co, and Fe) used to catalyze DRM, Ni based catalysts presented good activities and economic feasibility [28–31], currently considered one of the most promising catalysts in DRM reactions due to the high catalytic activity and low cost [32–38].

However, the recent progress of DRM catalyzed by DBD plasma on Ni-based catalysts is rarely summarized. Therefore, this article focuses on the application of DBD-Ni catalysts integrated system for DRM reaction, discusses the synergy between Nickel-based catalysts and DBD reactor, introduces the reactor design and process parameter optimization, and finally provides the current challenges and prospects for the future.

2. Overview of Non-Thermal Plasma

Plasma takes up around 99% of the universe substances, which is a term used to describe an ionized gas. It is a macroscopic appearance composed of free electrons, ions, and neutral particles (in which positive and negative charges are equal), forming an electrically neutral non-condensing system [39,40]. The plasma as a whole is not charged, but because it contains free charge carriers, it is conductive. In addition, it has strong chemical activity. Many chemically stable substances can be activated by plasma [2].

The classification of plasma is divided into thermal plasma (equilibrium plasma) and non-thermal plasma (non-equilibrium plasma) according to the temperature of heavy particles inside the plasma and the thermodynamic balance. In thermal plasma, the temperature of its electrons and heavy particles are approximately equal, between 5000 and 50,000 K; including arc discharge plasma and inductively coupled plasma, often used in solid waste incineration and arc welding [2,40]. The electron temperature in NTP is much higher than that of heavy particles. Therefore, NTP technology has a wide range of applications such as pollutant removal, nano-material synthesis, material surface modification, and fuel reforming [24,26].

2.1. Characteristics of Non-Thermal Plasma (NTP)

NTP has non-equilibrium characteristics, that is, it can simultaneously have higher electron energy and lower ion and gas temperature. Electrons with sufficiently high energy can activate reactant molecules by dissociation and ionization during collisions; also, the reaction system can be kept near room temperature, which can reduce the energy consumption. Therefore, non-thermal plasma has a wide range of applications [25,39].

The energy in the non-thermal plasma is mainly used to generate highly reactive species (such as methane and carbon dioxide). The non-equilibrium characteristics can overcome thermodynamic obstacles in chemical reactions (such as dry reforming), allowing the reaction to proceed at room temperature and pressure. Besides, the on-off switch time of non-thermal plasma is so short that the excess energy generated by the fluctuation of the power grid can be used to achieve the stability of the power grid and facilitate the control of the reaction [35]. Therefore, non-thermal plasma technology is promising for methane dry reforming.

2.2. Non-Thermal Plasma Generation

Plasma can be generated from neutral gas by thermal excitation. When the gas is strongly heated up to a certain temperature, usually thousands of kelvins, the gas molecules form plasma due to their enough energy for spontaneous dissociation, excitation and ionization. However, due to technical issues and high energy input, this plasma generation method is not popular [2].

In contract, the widely accepted and facile NTP generation method is by means of electricity. Capacitively coupled plasma is a typical example, where a large potential difference is exerted between two electrodes. Due to the electrical discharge generated by the electric field, the gas between the electrodes is transformed into plasma. Considering the thousands of volts voltage and short distance between electrodes, the intensity of the electric field in gas is high enough to accelerate the electrons from one electrode to another [2]. The current common non-thermal plasma mainly includes: Dielectric barrier discharge, corona discharge, gliding arc discharge, glow discharge, microwave discharge, and radio frequency discharge.

2.3. Mechanism of DRM Catalyzed by DBD Plasma

DBD is defined in the scenario where the dielectric is placed between two electrodes, and the discharge space is filled with an insulating medium. Due to the existence of the medium, the growth of the discharge current is limited, so as to avoid the complete breakdown of the gas and the formation of sparks or Arc [26]. When electricity is applied, plasma can be generated via the charge accumulation on the dielectric material with the short lifetime streamer, which will stop the discharge at one point, thus driving the discharge to happen at another point on the surface [41]. Moreover, DBD is a technology to generate stable plasma in a large scale at a low temperature and to produce electrons with a high energy to activate and dissociate greenhouse gas molecules such as CH_4 and CO_2, which are usually chemically stable due to the high bonding energies. On the other hand, dielectric barrier discharge can generate stable and uniform atmospheric pressure plasma under atmospheric pressure or higher. After optimization, this type of discharge has a broad development prospect in the industry.

DBD plasma is a means to effectively activate molecules. First of all, it can often activate high-stability methane molecules under milder conditions, usually at a pressure of 104–106 Pa and a frequency of 50 Hz [42]. Secondly, the simple design and easy-to-operate characteristics of the DBD reactor facilitates the miniaturization or expansion with high portability [39]. At the same time, DBD plasma shows certain advantages to DRM because of its low energy input, production of various active species, and low installation cost [18,43]. Therefore, this technology has been extensively studied in DRM [44].

Despite the advantages of DBD-catalytic assisted DRM over the single DRM, the low energy utilization efficiency limits its application [10,21,26]. To solve the problem, the combination of plasma technology and catalysts can improve the reactivity and energy utilization efficiency. It is known that the design of the plasma reactor should enable the plasma species optimally transported to the catalyst surface. Due to the high temperature of the catalyst, it is not simple to integrate the catalyst in the warm plasma; therefore, for plasma catalysis, DBD plasma as the representative NTP is more suitable [10,21].

2.4. Synergy of Non-Thermal Plasma and Catalyst

Although DRM shows a better performance in dielectric barrier plasma reactors, the traditional method to convert a large amount of CO_2 or CH_4 has many limitations. For example, other possible reactions, such as methane cracking, lead to reduced energy efficiency and residual carbon deposition [21]. In addition, the DRM process is highly endothermic with a long duration and high energy consumption. Therefore, the advantages of the combination of plasma and catalyst are exploited. Compared with the catalyst-free DRM, plasma-assisted catalytic DRM exhibited a higher reaction rate and strong resistance to coking due to the synergistic effect [45,46]. In detail, adding the catalyst to the discharge area of the DBD plasma can affect the efficiency of the system. Charge accumulation and polarization effect caused by the various shape and surface properties of catalysts enhance the electric field of plasma [8]. In other words, The plasma-catalyst interface increases the likelihood of reactant collisions and enhances surface modification and electric fields. These factors are beneficial to the treatment of gas in the DBD plasma DRM. On the other hand, plasma physically and chemically affects the surface area and interfacial chemistry of the catalyst, which in turn enhances DRM activity and improves product distribution [47].

The main factors affecting the plasma performance (measured in terms of output voltage and current intensity) are electric field enhancement and discharge. The roughness of the catalyst has a certain impact on it, due to the enhanced polarization effect and accumulation of charge in a more dispersed and smaller particle with high surface area and high density of metal sites [48]. The combination of the catalyst and the DBD plasma increases the reactivity of most reactants by keeping them in excited state and lowering the activation energy [8,49]. In a word, the synergy between catalysts and DBD plasma can be summarized based on Figure 1 as below: First, the carbon deposition is alleviated; second, the conversions, selectivity, and stability are enhanced; third, catalyst reusability is improved; fourth, energy efficiency is increased [8].

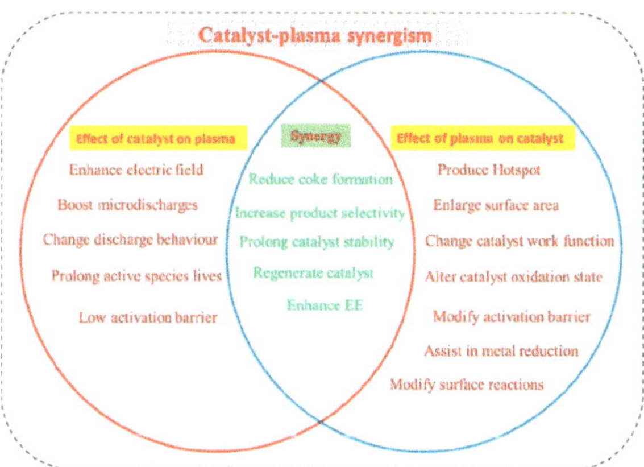

Figure 1. Synergy of the catalytic dielectric barrier discharge (DBD) plasma system. Reproduced with permission from [8], Copyright 2019, Elsevier.

In recent years, many studies have been focused on the application of a plasma synergistic catalyst to the DRM process [9,26,50]. A series of Co, Ru, Mn, and Pt-based catalysts have been published and most of them can increase the yield of products. Among them, Ni has become a hotspot due to the same stability, high activity, and high availability for DRM reactions as noble metals but with a lower cost [51]. For example, Wang et al. compared two DRM processes: Plasma-single and plasma-catalysis [26]. The results show that when Ni/C catalyst is combined with plasma, the catalytic activity was enhanced and the reaction conversion rate increased. Pietruszka et al. found that by combining plasma and Ni/γ-Al$_2$O$_3$, conversions of both CH$_4$ and CO$_2$ were increased due to the discharge heating of the catalyst [24]. Kim et al. observed that with Ni/Al$_2$O$_3$ catalyst, conversion of CH$_4$ was significantly higher than that of DBD without catalyst [45,50,52,53]. Therefore, this article summarizes the latest progress of Ni-based catalyst assisted by DBD plasma for DRM.

3. Ni-Based Catalyst Assisted with DBD Plasma for DRM

Due to the high availability and low cost of nickel-based catalysts, it is widely considered in plasma DRM. The common supports used in Ni-based catalysts are Al$_2$O$_3$, La$_2$O$_3$, SiO$_2$, AC, ZrO$_2$, and multi-element supports [8]. Studies have shown that different supports used in Ni-based catalysts exhibit different performances. For example, coking and metal sintering can be reduced in inert supports such as silica, thus improving the catalytic activity because CH$_4$ and CO$_2$ are both activated by metals with the help of plasma [8]. Besides, Ni-impregnated alumina (acid) and magnesium oxide (alkaline) catalysts have been widely used in plasma DRM, and their catalytic property has been improved by the dual-function approach, where CH$_4$ is activated by metal and CO$_2$ is activated by alkaline supports [54]. Specifically, alumina's high specific area, high thermal resistance, and good dielectric properties make it the preferred support for plasma catalysis [53]. In addition, Ni/ZrO$_2$ decomposed by plasma has a higher specific surface area, which generates more surface Ni active centers [55] (Table 1). Furthermore, activated carbon (AC) has the advantages of low cost, large specific surface area, good stability, chemical inertness, and has also been applied to DRM [56].

3.1. Pure Ni Catalysts with Different Supports

3.1.1. Ni/SiO$_2$

In the traditional DRM catalysis process, the sintering of nickel catalyst under high temperature reaction (generally above 500 °C) is one of its main disadvantages. The aggregation of nickel particles leads to the loss of active surface, thereby reducing activity and selectivity [57]. On one hand, DBD plasma and Ni/SiO$_2$ catalyst show a synergy, which inhibited coke formation. On the other hand, due to the higher energy of electrons in DBD than the bond dissociation energy of nickel, metallic Ni crystal grains are reduced to smaller crystal grains on the Ni/SiO$_2$ catalyst [58]. In the strong electric field of the DBD, the Ni-support interaction is affected by high-energy electrons, resulting in the dispersion of Ni particles on the catalyst support. In addition, DRM catalyzed by Ni/SiO$_2$ occurs at the low temperature and ambient pressure of non-thermal plasma, which may prevent nickel from sintering on the catalyst. As shown in Figure 2, a broader peak shape was presented after the reaction in 2b, suggesting a smaller size for the spent Ni catalyst due to the assistance of non-thermal plasma [59] (Table 1).

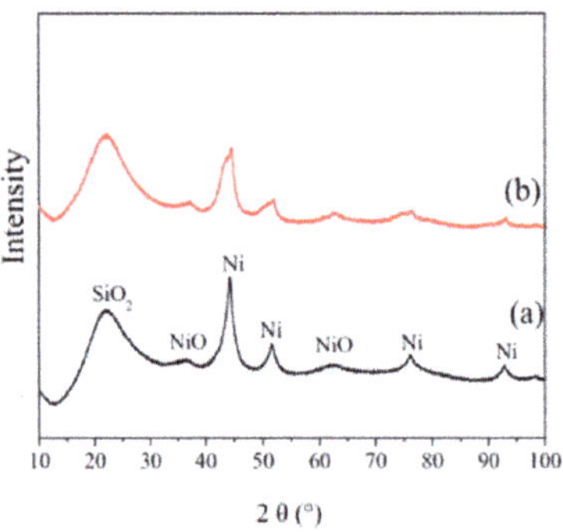

Figure 2. XRD pattern of Ni/SiO$_2$ catalyst sample. (**a**) Fresh catalyst. (**b**) Spent catalyst after 5 h test. Reproduced with permission from [59], Copyright 2015, Elsevier.

In addition, the Ni/SiO$_2$ catalyst can increase the CO selectivity, because the reaction between carbon-containing intermediates and oxygen radicals are catalyzed in the plasma. However, in the Ni/SiO$_2$ catalyst catalysis process, the conversion rate of CO$_2$ and CH$_4$ will be reduced, which can be attributed to the reverse reaction on the catalyst [59].

3.1.2. Ni/Al$_2$O$_3$

Due to its high thermal resistance, high specific area, and good dielectric properties, alumina was once known as the preferred carrier for plasma catalysis. Recent studies have shown that among Ni/MgO, Ni/γ-Al$_2$O$_3$, Ni/TiO$_2$, and Ni/SiO$_2$, Ni/γ-Al$_2$O$_3$ catalyst performed the best [60]. Nickel-supported Al$_2$O$_3$ has been widely used and reported in NTP-catalyzed DRM [54]. Tu and Whitehead reported that a synergistic effect of Ni/Al$_2$O$_3$ with low-temperature plasma almost doubled the methane conversion and H$_2$ yield compared with DRM with plasma alone [37,61].

L. Brune and coworkers also performed DBD plasma-DRM reactions with Ni/γ-Al$_2$O$_3$ as fillers [21]. Results showed that porous alumina with a huge specific area captured CO2 molecules, which significantly increased its conversion. Additionally, with the same wall plug power, total flow rate, and operating frequency, porous γ-Al$_2$O$_3$ beads filled in the gaps of the DBD increased the CH$_4$ conversion rate by nearly 25% compared with the plasma alone [21,62].

To enhance H_2/CO ratio in Ni/Al_2O_3 catalyst coupled with DBD reactor, Thitiporn et al. increased the CH_4/CO_2 feed ratio to 4 and a high H_2/CO ratio of 1.5 was realized; moreover, with 5% Ni loaded on Al_2O_3, only 3.7 wt% carbon deposition was presented, which was 4.7 wt% without Ni, suggesting the synergy between Ni catalyst and DBD [63] (Table 1). However, even though Ni/γ-Al_2O_3 exhibits good catalytic performance, there is still a problem of carbon deposition. The deposited carbon covers the active center of the catalyst and causes metal particles to sinter, thus deactivating the catalyst and reducing the DRM performance [43,64].

3.1.3. Ni/ZrO_2

ZrO_2 has been studied as the packing materials in DBD plasma reactor to catalyze the DRM reaction. However, with ZrO_2 alone, the fraction of void decreased, reducing the current intensity and plasma generation. Moreover, surface discharges became dominant instead of filamentary microdischarges, negatively affecting the activation of reactant molecules [23].

In contrast, Ni/ZrO_2 exhibits an enhanced reactivity in DBD DRM reaction. The nickel precursor is decomposed under the DBD plasma to prepare a nickel/zirconia catalyst, possessing a high specific surface area and provides more surface Ni active centers for the reaction. During the early work, the application of Ni/ZrO_2 catalysts in dry reforming has been studied, and it is found that carbon deposition is significantly reduced, especially under low nickel loading, if the nickel microcrystals used are very small [65]. ZrO_2 contains two crystal phases, namely monoclinic (m-ZrO_2) and tetragonal (t-ZrO_2). In ZrO_2-P (precursor decomposed with plasma treatment in Ar atmosphere), the amount of t-ZrO_2 is twice that of Ni/ZrO_2-C (precursor calcined in air without plasma treatment), indicating the beneficial effect of NTP treatment on the formation of t-ZrO_2 [55]. Compared with Ni/ZrO_2-C, the agglomeration of nickel particles on the Ni/ZrO_2-P was much reduced, and the exposed surface lattice fringes of the Ni particles were clearly visible. In addition, more oxygen vacancies were generated on Ni/ZrO_2-P, which provided stronger alkalinity and promoted CO_2 adsorption and activation, enhancing the activity of Ni/ZrO_2 in DRM. Meanwhile, oxygen vacancies additionally drive the CO_2 reduction, resulting in enhanced activation of CO_2 and Ni/ZrO_2 DRM activity [56,66]. The methane and CO_2 conversions using plasma and catalyst were 53.57% and 60.81%, higher than 42.30% and 52.88% with plasma alone, indicating that the combined treatment of DBD plasma and Ni/ZrO_2 can provide stronger adsorption sites and improve the adsorption capacity of CO_2, further increasing the activity of Ni/ZrO_2 in DRM. Finally, by comparing the carbon deposition amount of the two catalysts after the DRM reaction, Ni/ZrO_2-P possessed lower coke formation (The carbon deposition amount of Ni/ZrO_2-P and Ni/ZrO_2-C is 33 wt% and 64 wt%, respectively) [55].

In summary, studies by Vakili and Hu et al. [23,55] have confirmed that DBD plasma decomposition benefits the formation of highly active Ni-based DRM catalysts. Through plasma decomposition, Ni/ZrO_2-P possessed a smaller nickel size and denser Ni (111) planes. At the same time, it has higher dispersion, more t-ZrO_2 and more oxygen vacancies, which helps CO_2 adsorption and activation. Owing to the synergistic effect of non-thermal plasma and Ni/ZrO_2, the catalytic activity and anti-coking ability for DRM are effectively enhanced [55,65].

3.1.4. Ni/AC

Carbon materials such as activated carbon (AC) are characterized with low cost, good stability, large specific surface area, and chemical inertness [56]. In addition, activated carbon is effective for oxygen reduction, controlling the surface chemistry and pore size distribution. In recent years, many studies have been launched on this field. For example, Wang et al. studied DRM with plasma alone, catalyst alone, and catalyst-assisted plasma to determine the synergy [60]. In the case of catalytic mode, since DRM is an endothermic reaction, a higher operating temperature is required to achieve a high balance of CH_4 and

CO_2 conversion. Considering the low temperature of 270 °C, the conversions were limited. With the plasma alone, the CH_4 and CO_2 conversion were 51.5% and 42.0%, respectively. Compared with the reforming in the catalytic-single or plasma-single mode, due to the synergistic effect of plasma and catalyst, extremely high conversions were obtained of 64.6% and 65.7% for methane [60]. Based on the characterization results, higher porosity and larger specific surface area enhanced the activity and stability, suggesting a good application potential of the plasma catalytic DRM.

3.2. Ni-Based Catalysts with Doping

Due to the plasma-catalyst synergy, DRM activities are significantly increased. However, doping with other metals or compounds can effectively resist carbon formation and optimize the product distribution. Hao and Yashima showed that compared with Ni/Al_2O_3 or Rh/Al_2O_3 catalysts, Rh promoted Ni/Al_2O_3 exhibited a better performance [67]. Moreover, metals such as Co, K, Mg, Mn, La, and Ce have been used as modifiers recently. Their combinations with metal Ni produce a higher reactivity for DRM [20,25,54,61,68–71].

3.2.1. Transition Metals

Transition metals have drawn interest as a doping agent to promote the performance of Ni catalysts for DBD-catalyzed DRM reaction, such as Mn [20] and Co [68]. The Ni-Co/Al_2O_3-ZrO_2 catalyst with an amorphous structure can improve the active phase dispersion and enhance the metal-support interaction, which is more conducive to the activation of reactants in DRM than the Ni/Al_2O_3 catalyst. Nader Rahemi et al. proved that the synergistic effect of Ni-Co/Al_2O_3-ZrO_2 and plasma can achieve higher CH_4 and CO_2 conversions and H_2 and CO yields [68] (Table 1). The Ni-Co/Al_2O_3-ZrO_2 nanocatalyst treated under 1000 V presented a uniform morphology, large surface area, and small particle size of 21.2 nm averagely, suggesting the great effect of plasma voltage on the crystallinity and size of NiO. In detail, the amorphous active phase of Ni-Co/Al_2O_3-ZrO_2 was easily attached to the crystal lattice of the support. The plasma improved the irregularity of the structure by generating kinks, vacancies, and other structural defects, causing the crystal grains to lose the lattice arrangement, thereby improving the active phase dispersion and strengthening the metal-support interaction. In the meantime, the fluid state time of the catalyst showed that the Ni-Co/Al_2O_3-ZrO_2 catalyst combined with plasma improved CH_4 and CO_2 conversions and the H_2/CO ratio compared with the traditional DRM [68].

Typical results suggest excellent activity and syngas yield under plasma conditions in Ni-Co/Al_2O_3-ZrO_2, however, the anti-carbon deposition and product selectivity of Al_2O_3-ZrO_2 catalyst can be improved. In this case, because Mn positively affects surface modification and activation of the catalyst, the emergence of Ni-Mn/Al_2O_3 catalyst causes widespread attentions. In Xintu's work, Ni/Al_2O_3 and Ni-Mn/Al_2O_3 catalysts were compared with plasma discharge [20] (Table 1). The experimental results showed that compared with plasma alone, catalytic DBD exhibited a higher conversion and syngas ratio. In detail, the Ni/Al_2O_3 catalyst possessed a strong affinity for CH_4, which is conducive to CO disproportionation, resulting in coke formation. Conversely, the Ni-Mn catalyst delivered a high conversion and stable activity. Under similar conditions, the amount of carbon produced on the surface of the Ni/Al_2O_3 and Ni-Mn/Al_2O_3 catalysts was 4.3 mg and 1.6 mg, respectively, which confirmed that the bimetallic catalyst Ni-Mn/Al_2O_3 was better than Ni/Al_2O_3. Moreover, a higher energy efficiency in DBD plasma-DRM was presented that when the discharge power was 1.0 W, the maximum energy efficiency of the bimetallic catalyst filled DBD was 2.76 mmol/kJ [20].

In general, DBD plasma reactor is well applied in DRM to produce hydrogen/syngas. Adding a catalyst to the DBD reactor can increase the conversions of reactants. The catalytic activity for CH_4 conversion is in order: Ni-Mn/Al_2O_3 > Ni/Al_2O_3 > Plasma alone. The best activity of the DBD filled with bimetallic catalyst is related to its resistance to carbon formation [20,72].

3.2.2. Alkali and Alkaline Earth Metals

In addition to transition metals, the addition of promoters such as alkali and alkaline earth metals (Mg and K [60,61]) can also inhibit the carbon formation, improve the dispersion of Ni metal, prevent metal sintering, strengthen the metal-support interaction thus enhancing the catalytic performance. One of the reasons is possibly their addition increases the alkalinity of the catalyst, thereby promoting the activation of CO_2 and CH_4 [31,65,73]. Recently, DeBeck and his colleagues discovered that hydrotalcite-derived catalysts (Ni-Mg/Al_2O_3) inhibited the Ni sintering and enhanced CH_4 conversion [74]. Tao et al. also doped Mg into the Ni/Al_2O_3 to enhance the CO_2 adsorption with the surface hydroxyl group, thus reducing the coke formation [75]. Sengupta et al. found that adding 5wt.% MgO to the Ni/Al_2O_3 catalyst can significantly reduce the carbon formation on the spent catalyst from 24.5 wt% to 14.4 wt% [76]. Ozkara-Aydinoglu et al. successfully reduced the carbon deposition from 4% to 2.6% when using Mg promoter in Co/ZrO_2 catalyst [77].

The mechanism can be summarized as below: the addition of Mg promoter can increase the number of strongly basic sites on the catalyst, thus significantly improving CO_2 adsorption. Coupled with DBD plasma, Ni-Mg/Al_2O_3 catalyst activated CO_2, but inhibited CH_4 conversion, thus reducing the carbon deposits in the product. On the TPR spectrum of the Ni-Mg/Al_2O_3 catalyst as shown in Figure 3, compared with the Ni/Al_2O_3 catalyst, the reduction temperature of NiO increases and the peak intensity related to the reduction of NiO decreases, suggesting that due to the plasma catalytic system and Mg promoters, the interaction between NiO and Al_2O_3 was greatly enhanced [61] (Table 1).

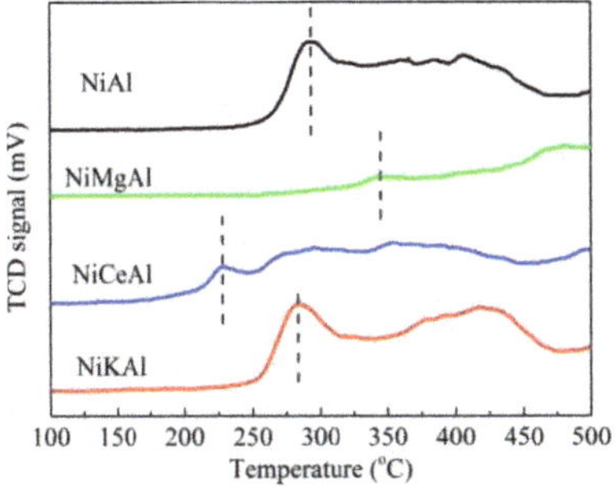

Figure 3. H_2-TPR spectrum of fresh catalyst. Reproduced with permission from [61], Copyright 2018, Elsevier.

Different from Mg, the addition of K the Ni-K/Al_2O_3 catalyst increases the conversions of the reactants and the yield of the products, and the energy efficiency of the plasma process is also greatly improved, which highlights the synergy between the plasma and the catalyst [60,61]. The interaction between DBD plasma and catalyst is the main driving force. Compared with plasma alone, integrating plasma with Ni-K/Al_2O_3 catalyst greatly increased CH_4 and CO_2 conversion to 31.6% and 22.8%, respectively. Moreover, H_2 selectivity was enhanced to 43.3% [61]. This is because the presence of K can increase the basicity of the DBD plasma catalytic system, which is beneficial to the conversion of carbon dioxide. Additionally, based on the TPR profiles, the reduction temperature of NiO in Ni-K/Al_2O_3 is significantly lowered, suggesting the formation of abundant active sites. However, compared with the Ni/Al_2O_3 catalyst, K doping increased the carbon deposition on the catalyst, consistent with the higher CH_4 conversions in the plasma catalytic system [60,61].

3.2.3. Rare Earth Metals

In the plasma-catalyzed DRM reaction, the Ni-K/Al_2O_3 catalyst is used to achieve a good yield of H_2 and CO. Similarly, rare earth metals such as La and Ce can act as doping agents to enhance the catalytic performance of Ni-based catalysts in DBD reactor for DRM [54,71]. Ce has been widely applied to improve the stability of DRM catalysts by inhibiting carbon formation and participating in the gasification of carbon deposits that have already formed [70,78–80]. For example, Ni-Ce/Al_2O_3 catalyst can effectively enhance the reactivity with the combination of plasma [70]. Wang et al. studied the catalytic DRM on NiO-CeO_2-Al_2O_3 catalyst and found that by increasing the ratio Ce/Al in Ni-Ce/Al_2O_3 from 0 to 1:50 resulted in higher conversions of CH_4 and CO_2. The presence of Ce further improves the catalytic performance of DRM under plasma conditions since CeO_2 has a higher dielectric constant, thus exhibiting a better dielectric performance [79]. Moreover, compared with the Ni/Al_2O_3 catalyst, Ce addition reduced the CO_2 desorption temperature by 20 °C, indicating that the combination of DBD plasma and the catalyst can weaken the CO_2 desorption at strong basic sites owing to the enhanced activation of CO_2 on CeO_2 species [61,70].

In addition to enhancing the catalytic performance, combination of Ni-Ce catalysts and DBD can prevent the coke formation. Tao et al. found that Ce addition would bring oxygen vacancies and promote the adsorption of CO_2, thus facilitating the coke removal. Moreover, the high specific surface area of the catalyst changed the filament discharge alone to filament/surface discharge mode, increasing the discharge area [75] (Table 1). Besides, in the DBD-treated Ni/SiO_2 catalyst, the interfacial interaction was obviously improved, which promoted the activation and dissociation of methane. Additionally, the Ni/CeO_2-SiO_2 catalyst decomposed by DBD plasma showed smaller Ni particle size and fewer defect centers, beneficial to improve the activity, anti-coking, and anti-sintering performance in the DRM reaction.

To investigate the Ce doping effect in Ni/SiO_2 with DBD, Yan et al. compared Ni/CeO_2-SiO_2-C (no plasma treatment) and Ni/CeO_2-SiO_2-P (DBD plasma treatment) for DRM [71]. The experimental results showed how the interaction between Ni and CeO_2 improved the catalytic performance. At 700 °C, Ni/CeO_2-SiO_2-P delivered a higher activity and H_2/CO ratio than Ni/CeO_2-SiO_2-C. The conversions of CO_2 and CH_4 were 87.3% and 78.5%, higher than the latter (80.5% and 67.8%). At the same time, Ni/CeO_2-SiO_2-P is stable in long-term reaction, while Ni/CeO_2-SiO_2-C shows poor stability within 10h [71].

The main difference between the two catalysts lied in the amount of active oxygen in the interface structure between Ni and CeO_2. Ni/CeO_2-SiO_2-P has more active oxygen than Ni/CeO_2-SiO_2-C. As shown in Figure 4a,b, a higher ratio of Ce^{3+}/Ce^{4+} of 0.24 was achieved in Ni/CeO_2-SiO_2-P than that of 0.18 in Ni/CeO_2-SiO_2-C, which suggested formation of more surface oxygen species. Additionally, Ni/CeO_2-SiO_2-P in Figure 4c presented a higher ratio of O_α/O_β where O_α referred to oxygen ions and O_β belonged to the surface hydroxyl groups, indicating more active oxygen generated with the plasma treatment. Therefore, Ni/CeO_2-SiO_2-P exhibited a better catalytic performance and higher H_2/CO ratio at lower than 700 °C. Based on the kinetic study, reactants were more easily activated in Ni/CeO_2-SiO_2-P than Ni/CeO_2-SiO_2-C, which confirmed the higher reforming activity of the former. In general, Ni/CeO_2-SiO_2 with plasma exhibited a good synergistic effect in DRM [71].

In addition to CeO_2, another rare earth metal oxide, La_2O_3 is proven effective as a doping agent. For example, $MgAl_2O_4$ structure facilitates the interaction and dispersion of the metal and changes the acidity and alkalinity of the catalyst [54]. In order to further increase the H_2/CO ratio, studies have shown that adding La_2O_3 can inhibit the formation of water [54] (Table 1). Besides, La_2O_3 used for the dry reforming of methane shows significant resistance to carbon formation due to strong alkalinity and carbon gasification [81]. La_2O_3 inhibits carbon deposition and prolongs the stability of the catalyst because it can form an intermediate carbonate ($La_2O_2CO_3$), which further reacts with the carbon on the surface near nickel to form CO_2. The interaction of nickel and magnesium aluminoxane can also be enhanced by the assistance of La_2O_3 as an additive. Under plasma conditions, mixed support La_2O_3 and $MgAl_2O_4$ can increase the alkalinity of the catalyst, inhibit the aggregation of nickel particles, and reduce the generation of carbon [54,82] (Table 1). Owing to the merits, 10% nickel/La_2O_3-$MgAl_2O_4$ prepared in DBD plasma possessed the highest CH_4 conversion (86%) and CO_2 conversion (84.5%), the highest H_2/CO ratio and the lowest carbon formation rate [54].

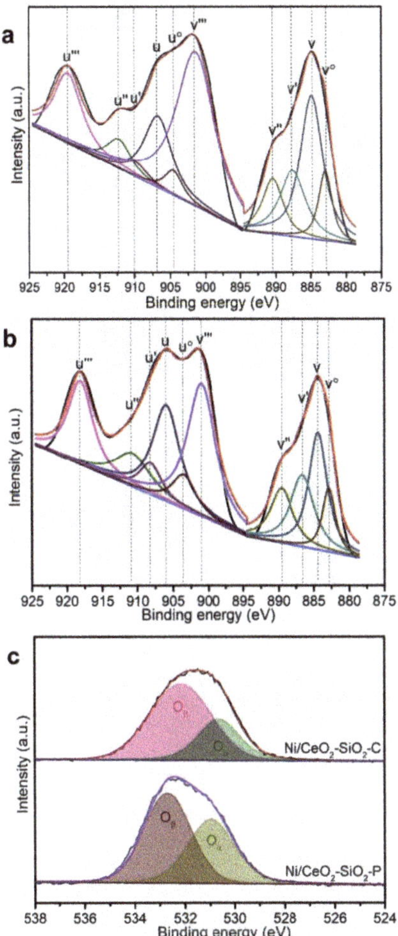

Figure 4. Ce 3d XPS spectra of (**a**) NiO/CeO_2-SiO_2-C and (**b**) Ni/CeO_2-SiO_2-P, and (**c**) O 1s XPS spectra of the two catalysts. Reproduced with permission from [71], Copyright 2019, Elsevier.

Finally, under similar experimental conditions, a comparative study of the synergistic effect of plasma and catalyst was carried out. The CH_4 and CO_2 conversions in the catalytic DBD plasma were 26% higher than those of the plasma alone. Due to the basic nature of the catalyst, the reactant gases were adsorbed on the surface of the catalyst, and more reactive substances diffused in the pores, allowing subsequent surface reactions to occur. When the reactant gas molecules collided with high-energy electrons, the electrons in the molecules were excited, dissociated, and ionized to produce active species such as O, H, methyl radicals, and finally CO and H_2. Therefore, the integration of Ni/La_2O_3-$MgAl_2O_4$ with the plasma facilitated the processing of the reactants and improved the efficiency of the DBD plasma reactor [54,83] (Table 1). In general, the preparation of the Ni/La_2O_3-$MgAl_2O_4$ improved the alkalinity of the catalyst, the Ni dispersion, the dielectric constant, and the interactions. Owing to these merits, the activation and chemical adsorption of reactants were enhanced.

Table 1. Catalytic DRM activities of Ni-based catalysts integrated with DBD plasma.

Catalyst	Parameters				Conversion(%)		Selectivity(%)		T (°C)	SIE * (J/mL)	Energy Efficiency (mmol/kJ)	Remarks	Ref.
	CH_4/CO_2 Ratio	Catalyst Loading (g)	Flow Rate (mL/min)	Power (W)	CH_4	CO_2	H_2	CO					
$LaNiO_3@SiO_2$	1:1	0.2	40	150	88.31	77.76	83.65	92.43	200	225	0.17	Integrated with DBD, $LaNiO_3@SiO_2$ shows improved conversions and selectivity.	[82]
Ni/SiO_2	1:1	0.2	50	86	26	16	47.4	52.9	110	103.2	N.A.	The combination of DBD and Ni/SiO_2 catalyst enhance the activity of DRM due to the reaction between carbon-containing intermediates and oxygen radicals.	[59]
Ni/Al_2O_3	1:1	0.3	56	70	60	77	70–75	80	550	75	39%	The charge recombination on the Ni/Al_2O_3 catalyst surface will enhance the diffusion of carbon through the Ni catalyst and promote its oxidation by CO_2.	[49]
Ni/Al_2O_3	4:1	6.4	50	1600	8.3	7.6	69	20	Ambient temperature	4.6 eV/molecule	4.5%	Ni addition enhanced the H_2/CO ratio and reduced coke formation with the help of DBD.	[63]
$Ni/\gamma-Al_2O_3$	1:1	1.0	50	50	56.4	30.2	31	52.4	<150	60	0.32	The combination of plasma and $Ni/\gamma-Al_2O_3$ can increase the conversion rate of CH_4.	[27]
$Ni/La_2O_3-MgAl_2O_4$	1:1	0.5	20	100	86	84.5	50	49.5	350	300	0.13	La_2O_3 inhibits the RWGS * reaction, improves H_2 selectivity and yield, and the formed intermediate carbonate ($La_2O_2CO_3$) inhibits carbon deposition.	[54]
Ni/La_2O_3	1:1	0.2	50	160	63	54	71	85	150	240	0.14	Ni/LaO_3 nanoparticles show excellent thermal stability in the DBD plasma reactor. La contributes to the formation of intermediates, which are responsible for activating CO_2 and inhibiting carbon deposition.	[83]
Ni/ZrO_2	1:1	0.6	50	200	53.57	60.81	82	95	650	240	N.A.	The Ni/ZrO_2 catalyst prepared by the DBD plasma decomposition method greatly improves its activity due to its high dispersion and increased oxygen vacancies.	[55]
$Ni-Co/Al_2O_3-ZrO_2$	1:1	0.3	40	N.A.	58	62	95	100	850	N.A.	N.A.	The $Ni-Co/Al_2O_3-ZrO_2$ catalyst after plasma treatment shows high catalytic activity due to its narrow particle size distribution, large surface area and strong metal-support interaction.	[68]
$Ni-Mn/\gamma-Al_2O_3$	1:1	0.5	30	2.2	28.4	13.2	23.2	40.5	N.A.	4.2	2.76	A higher activity and energy efficiency is achieved by the integrated plasma and Ni-Mn bimetallic catalyst system.	[20]
$Ni-Mg/Al_2O_3$	1.6:1	0.4	50	16	32	16	41.7	29.5	160	19.2	0.58	K promoted catalyst shows the best performance and enhances the energy efficiency of plasma process because it contains more active sites.	[61]
$Ni-K/Al_2O_3$	1.6:1	0.4	50	16	34	23	43.3	31.3	160	19.2	0.67		
$Ni-Ce/Al_2O_3$	1.6:1	0.4	50	16	32	22	41.8	31.1	160	19.2	0.63		
NiMgAlCe	4:6	0.45	90	48	36.1	22.5	N.A.	N.A.	N.A.	32	N.A.	Mg and Ce promoted the CO_2 adsorption and increased the discharge area by partially tuning the filament discharge into surface discharge.	[75]
$Mg,Ce-Ni/\gamma-Al_2O_3$	1:1	N.A.	30	2.7	34.7	13	35	53.7	N.A.	5.4	1.97	Mobile oxygen and surface basicity effectively removed coke during methane activation by Ni and DBD.	[80]

* RWGS stands for the reverse water gas shift reaction; * SIE refers to specific input energy; N.A.: not available.

4. DBD Plasma Reactor Design

To achieve a high catalytic performance and energy efficiency, reactor designs including the configuration of the reactor, medium materials, and discharge volume play an essential role. In the following part, DBD plasma reactor designs will be illustrated in detail.

4.1. Configuration of DBD Plasma Reactor

Commonly used reactor configurations in DBD plasma include planar and cylindrical types. Typical planar DBDs reactor is composed of two electrodes (as shown in Figure 5), asymmetrically located on top and bottom sides of the dielectric barrier material, which limits the current and inhibits the sparks formation. Electric discharge is generated when an AC potential is exerted [24,26,63].

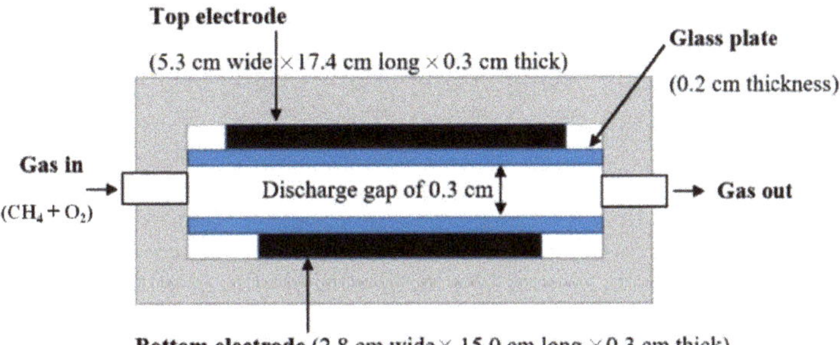

Figure 5. Two frosted glass plates are used to mix parallel plate DBD reactors. Reproduced with permission from [63], Copyright 2020, Springer.

Different from the planar DBD plasma reactor, as shown in Figure 6a, the cylindrical plasma reactor contains the following geometric parameters, which are discharge volume (VD, cm^3), the discharge length (DL, cm), discharge gap (D_{gap}, mm), and shape and material of the high-voltage electrode [8,41]. In the basic structure of the DBD reaction device, the high-voltage electrode is usually a rod-shaped material with high conductivity similar to steel and aluminum [84]. Generally, high-voltage electrode materials have three morphologies as shown in Figure 6b, including holes, porous, and rough structures [8]. The morphology of the high-voltage electrode is a very important factor in a DBD reaction device, usually determining whether the discharge is uniform. The uniform distribution of the high-voltage electrode can make the gas molecules and the energy-containing material fully interact, which improves the reaction rate of the DBD plasma in the reaction device. Moreover, researchers have studied the high-voltage electrode material and its surface morphology [22]. It is concluded that the surface conductivity can be increased by changing the high-voltage electrode surface morphology. The usual method is to deposit zinc oxide on the surface of the high-voltage electrode, thereby producing a more uniform surface and a stable plasma discharge [85]. On the other hand, the morphology will cause the discharge mechanism to be different, thus affecting the conversions of methane and carbon dioxide. With a porous electrode, a higher conversion will be obtained than that of a smooth high-voltage electrode. The reason is that the gas molecule fixation rate in the porous electrode is higher, thereby increasing the collisions of the active species DRM.

In addition to the rough morphology, uniformly coating the catalyst on the rough high-voltage electrode surface, as shown in Figure 6c, makes the catalyst active center more direct contact with the plasma [8]. Due to the robust discharge behavior and uniformity of the discharge, it can also produce reactive species. However, one drawback is carbon nanotube (CNT) growth and surface etching.

Besides the shape and morphology, the volume of the high-voltage electrode also greatly affects the discharge characteristics of the DBD plasma [8]. When reactant molecules

(CH$_4$, CO$_2$) act in the reactor with DBD plasma, the volume of the high-voltage electrode affects the discharge volume, thereby influencing the conversion rate of CH$_4$ and CO$_2$.

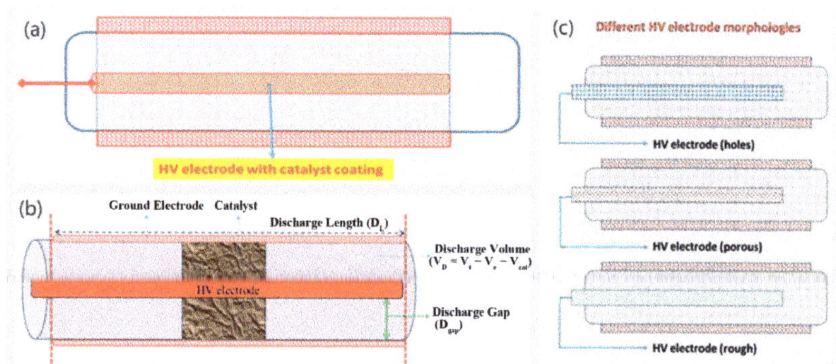

Figure 6. (**a**) The typical geometry of a DBD plasma reactor system; (**b**) the morphology of three different high-voltage electrode materials; (**c**) three morphologies of high-voltage electrode materials. Reproduced with permission from [8], Copyright 2019, Elsevier.

As one of the representative DBD reactors for plasma-assisted catalysis, packed-bed reactors with different designs are shown in Figure 7 [8]. Pre-packing mode is mostly applied in those temperature-controlled reactions. The catalysts needs to be activated first, followed by passing the reactants to the discharge zone, which may not be practical in DBD-DRM system. In contrast, in situ packing reactor takes advantages of the catalyst and plasma simultaneously. With a relatively stable temperature, the excited electrons promote the dissociation of reactant molecules. Few researchers focus on the post packing and fully packed DBD reactor in DRM reaction since they might be not effective as the in situ packing mode.

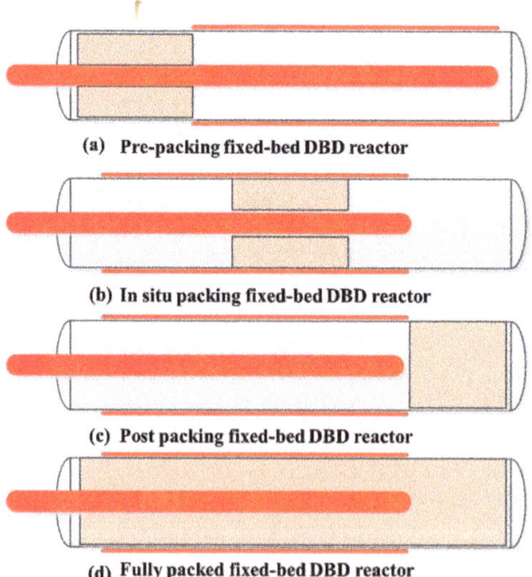

Figure 7. Different reactor configurations based on the packing material. (**a**) Pre-packing fixed-bed DBD reactor; (**b**) in situ packing fixed-bed DBD reactor; (**c**) post packing fixed-bed DBD reactor; (**d**) fully packed fixed-bed DBD reactor. Reproduced with permissionfrom [8], Copyright 2019, Elsevier.

4.2. Medium Material

The main function of the dielectric material in the DBD reactor is to separate the two electrodes of the reactor, namely the high-voltage and ground electrode; also, via adsorption of the discharge species, the discharge voltage is lowered. The dielectric material is mainly important for the mechanism of the discharge, because excitation, ionization, and dissociation are directly related to the dielectric constant in the plasma-induced reaction. The material with high dielectric constant can increase the temperature of the reactor by reducing the power consumption and thus increasing the electric field. We can increase the dielectric constant to improve the activity of the reaction and increase the temperature of the reactor [86,87].

In DBD plasma, different dielectric materials can often be used as dielectric barriers. Quartz and alumina are two representative dielectric materials affecting the discharge behavior of DBD plasma and the dissociation of CH_4 [86,88]. Compare with quartz, alumina has a higher dielectric constant and porous morphology. At the same time, the dielectric material also affects the discharge behavior with respect to the applied voltage and discharge current. Due to the high discharge power of the reactor in alumina, the high electric field generated by its partial porous structure helps to activate the reactant species in the pores of the alumina material.

4.3. Discharge Volume

The V_D of the DBD plasma reactor is composed of D_{gap} and DL [89]. D_{gap} can be expressed by using the difference between the inner radius of the dielectric and the outer radius of the high-voltage electrode. Duan et al. investigated the effect of D_{gap} on CO_2 conversion in DRM reaction and found that a larger D_{gap} could increase the CO_2 conversion due to the increased residence time of CO_2, leading to a prolonged contact with other reactive species [90]. In another study, 3 mm D_{gap} realized the highest conversion of CO_2 and methane when D_{gap} varied from 1–5 mm, suggesting the necessity of optimal selection of an appropriate D_{gap} value [87].

DL is another geometric parameter in DBD plasma. The length of the electrode can affect the value of DL and corresponding VD. Relevant studies suggested that increasing DL can increase the conversion rate, which is due to the higher contact time of the reactive species and gas molecules at a constant flow rate [87].

5. Effects of Process Parameter

Process parameters exert crucial influences on realizing the industrialization of technology. Many studies have described the process parameters, such as input power (Pin), feed flow rate, specific input energy (SIE), and feed ratio [69,91,92]. These parameters have a certain impact on gas processing and system efficiency. Therefore, by adjusting these process parameters, it can help to allocate targeted products reasonably.

5.1. Input Power

Input power is an important process parameter of catalytic DRM reaction by DBD plasma. It is related to the energy efficiency (EE) of the gas treatment and processing. Generally speaking, for a DBD plasma reactor, as the input power increases, the CH_4 and CO_2 conversion rate increases. This is because the increased input power can increase the electron density, accelerate the collision of the reaction gas molecules with high-energy electrons, and promote the activation of the reactants. These excited, dissociated, and ionized reactant molecules triggers the dry reforming reaction of methane [89].

The discharge power can be adjusted by adjusting the voltage (V) and frequency (f). As the voltage and frequency increase, the current pulse increases, thereby increasing the processing of the reactant gas. For example, Liu et al. found that the selectivity of H_2 and CO in DBD increases with the increase of discharge power [69]. On the other hand, the frequency can change the reactivity within a certain range; once this range is exceeded, the conversion rate will no longer be affected by the frequency [87,93]. A similar phenomenon can also be seen at a voltage proportional to the plasma discharge power.

5.2. The Feed Flow Rate

The feed flow rate is another vital process parameter in the DBD plasma reactor, considering the impact on the conversion of the CO_2 and CH_4. Many studies have shown that various parameters are related to the total flow rate of the feed [94]. For example, Eliasson et al. studied the feed flow range of 200 to 900 mL min^{-1} and found that CO_2 and CH_4 conversions dropped from 54% and 60% to 18% and 23%, respectively [8]. Rico and colleagues found that with the increase of flow rate, CH_4 and CO_2 conversions decreased. However, the selectivity of H_2 and CO is not affected despite the highly affected yield.

The flow rate of feed is also linked to the mass transfer limitation and processing capacity of the DBD plasma reactor, and is inversely proportional to the reactants' residence time in the discharge zone [87]. The processing capacity of the reactor can be improved by adding a potential catalyst in DBD plasma. Mass transfer limitations can be overcome by high flow rates. At higher flow rates, the external mass transfer limitation will be minimized, but high flow rates will also cause some adverse effects [93,95].

5.3. Specific Input Energy

Specific input energy (SIE) refers to the energy required for processing raw gas into products or the energy required per liter of raw gases. SIE can be tuned by changing the input power and the total gas feed flow rate. Pinhao et al. studied DBD plasma methane dry reforming and found that the flow rate of feed and SIE are inversely proportional to each other. Therefore, CH_4 and CO_2 conversions are directly related to SIE [91]. With feed flow rate unchanged, the SIE increases with the increase of the input power, which in turn leads to an increase in catalytic performances, while the system efficiency of the DBD plasma reactor decreases [59,96].

5.4. The Feed Ratio

The feed ratio affects the selectivity H_2. Figure 8 presented the CO_2/CH_4 molar ratio effect on the DRM performance in a DBD plasma reactor without catalyst packing. In Figure 8, as the CO_2/CH_4 ratio increased, the CH_4 conversion and CO selectivity increased almost linearly. When the CO_2/CH_4 ratio increased from 1:9 to 1:3, the CO_2 conversion dropped from 41.5% to 21.3%; however, when further increasing the ratio to 9:1, the CO_2 conversion slightly increased as shown in Figure 8a. In addition, by increasing the CO_2/CH_4 ratio further to 9:1, the H_2 selectivity was close to 100%, and the CO selectivity increased by 8.8 times, while the C_2H_6 selectivity decreased by 53% (Figure 8b) [27] (Table 1). These studies paved the way to optimize the conversion and selectively by adjusting the feed ratio.

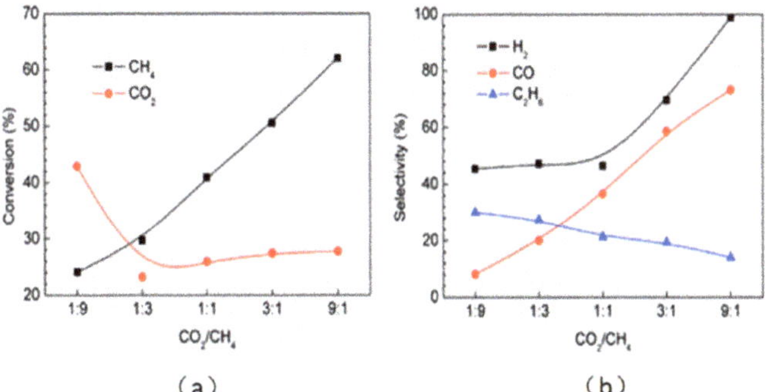

Figure 8. The CO_2/CH_4 molar ratio effect on the catalytic performances of plasma dry reforming of methane (DRM) reaction: (**a**) Conversion; (**b**) selectivity. (Total feed flow 25 mL min^{-1}, discharge power 15 W). Reproduced with permission from [27], Copyright 2015, Elsevier.

6. Conclusions and Outlook

This article reviews the recent development of DRM catalyzed in DBD plasma reactors on Ni-based catalysts, including an overview of non-thermal plasma, the mechanism of DRM with different plasmas, the synergy of plasma and Ni-based catalyst, reaction design, and process parameter optimization. The synergistic effect of DBD plasma and different Ni catalysts are classified and discussed in two categories: Pure Ni with various supports and Ni-based catalysts with doping. The synergistic effects of DBD plasma and pure Ni catalyst include size reduction of Ni grains, promoted discharge heating, higher efficiency of the DBD plasma reactor, and enhanced reactant adsorption. By doping with other metals and oxides, a higher active phase dispersion and stronger metal-support interaction are realized; the alkalinity of the catalyst is increased to promote the activation of CO_2 and CH_4; due to the higher dielectric constant, the dielectric performance of the plasma is better. Although great progress has been made in this area, the technology still faces many challenges, and there are still areas for further improvement:

(1) For DBD plasma-catalytic DRM, one challenge is power dissipation and carbon deposition. The trend of coke formation can be observed through the carbon balance and H_2/CO ratio [97]. The H_2/CO ratio close to unit 1 is considered to be an ideal method for processing syngas, and a lower carbon balance suggests a higher coke deposition. Carbon deposition in DRM is not only due to side reactions, but also due to the feed ratio of CH_4/CO_2 and other parameters [21]. Although catalysts have been used to avoid the formation of excess carbon and increase the reaction rate, the problem of carbon deposition has not been completely eliminated. Therefore, we continue to explore solutions to achieve better selectivity and stability.

(2) The synergy between the plasma and catalyst has been confirmed in recent years, but the interaction between the two has not been fully explored. Advanced methods are needed to study the mechanisms in depth to realize the rational design of the catalyst-plasma system [98].

(3) Efficient dry reforming of methane provides excellent energy efficiency, which is worthy of further consideration and research. In order to industrialize plasma catalysis, it is necessary to have a deeper understanding of the relationship between the catalytic performance and energy efficiency vs. reactor configuration and process parameter.

Author Contributions: X.G.: Conceptualization, data curation, investigation, writing—original draft, writing—review and editing; Z.L.: Conceptualization, data curation, investigation, writing—original draft, writing—review and editing; T.L.: Writing—original draft; L.H.: Writing—original draft; J.Z.: Writing—review and editing; S.A.: Data curation, writing—review and editing; N.D.: Data curation, writing—review and editing; A.J.: Project administration, supervision, validation; S.K.: Funding acquisition, resources, project administration, supervision, validation. All authors have read and agreed to the published version of the manuscript.

Funding: This research was funded by Ministry of Education in Singapore (MOE) Tier 2 grant (WBS: R279-000-544-112), Singapore Agency for Science, Technology and Research (A*STAR) AME IRG grant (No. A1783c0016), National Environment Agency (NEA) in Singapore (WTE-CRP 1501-103) and Youth Innovation Talents Project of Guangdong Universities (natural science) in China (2019KQNCX098).

Data Availability Statement: All data included in this study are available upon the permission from the publishers.

Conflicts of Interest: The authors declare no conflict of interest.

References

1. Snoeckx, R.; Zeng, Y.X.; Tu, X.; Bogaerts, A. Plasma-based dry reforming: Improving the conversion and energy efficiency in a dielectric barrier discharge. *RSC Adv.* **2015**, *5*, 29799–29808. [CrossRef]
2. Harinarayanan, P.; Damjan, L.J.; Dasireddy, V.D.B.C.; Likozar, B. A review of plasma-assisted catalytic conversion of gaseous carbon dioxide and methane into value-added platform chemicals and fuels. *RSC Adv.* **2018**, *8*, 27481–27508.
3. Gao, X.; Liu, H.; Hidajat, K.; Kawi, S. Anti-Coking Ni/SiO2Catalyst for Dry Reforming of Methane: Role of Oleylamine/Oleic Acid Organic Pair. *ChemCatChem* **2015**, *7*, 4188–4196. [CrossRef]
4. Kathiraser, Y.; Thitsartarn, W.; Sutthiumporn, K.; Kawi, S. Inverse NiAl2O4 on LaAlO3–Al2O3: Unique Catalytic Structure for Stable CO2 Reforming of Methane. *J. Phys. Chem. C* **2013**, *117*, 8120–8130. [CrossRef]
5. Gao, X.; Hidajat, K.; Kawi, S. Facile synthesis of Ni/SiO 2 catalyst by sequential hydrogen/air treatment: A superior anti-coking catalyst for dry reforming of methane. *J. CO2 Util.* **2016**, *15*, 146–153. [CrossRef]
6. Kawi, S.; Kathiraser, Y.; Ni, J.; Oemar, U.; Li, Z.; Saw, E.T. Progress in Synthesis of Highly Active and Stable Nickel-Based Catalysts for Carbon Dioxide Reforming of Methane. *ChemSusChem* **2015**, *8*, 3556–3575. [CrossRef] [PubMed]
7. Gao, X.; Ashok, J.; Widjaja, S.; Hidajat, K.; Kawi, S. Ni/SiO2 catalyst prepared via Ni-aliphatic amine complexation for dry reforming of methane: Effect of carbon chain number and amine concentration. *Appl. Catal. A Gen.* **2015**, *503*, 34–42. [CrossRef]
8. Khoja, A.H.; Tahir, M.; Amin, N.A.S. Recent developments in non-thermal catalytic DBD plasma reactor for dry reforming of methane. *Energy Convers. Manag.* **2019**, *183*, 529–560. [CrossRef]
9. Chung, W.-C.; Chang, M.-B. Review of catalysis and plasma performance on dry reforming of CH4 and possible synergistic effects. *Renew. Sustain. Energy Rev.* **2016**, *62*, 13–31. [CrossRef]
10. Uytdenhouwen, Y.; Hereijgers, J.; Breugelmans, T.; Cool, P.; Bogaerts, A. How gas flow design can influence the performance of a DBD plasma reactor for dry reforming of methane. *Chem. Eng. J.* **2021**, *405*, 6618. [CrossRef]
11. Bian, Z.; Das, S.; Wai, M.H.; Hongmanorom, P.; Kawi, S. A Review on Bimetallic Nickel-Based Catalysts for CO2 Reforming of Methane. *ChemPhysChem* **2017**, *18*, 3117–3134. [CrossRef] [PubMed]
12. Li, Z.; Li, M.; Bian, Z.; Kathiraser, Y.; Kawi, S. Design of highly stable and selective core/yolk–shell nanocatalysts—A review. *Appl. Catal. B: Environ.* **2016**, *188*, 324–341. [CrossRef]
13. Kathiraser, Y.; Oemar, U.; Saw, E.T.; Li, Z.; Kawi, S. Kinetic and mechanistic aspects for CO2 reforming of methane over Ni based catalysts. *Chem. Eng. J.* **2015**, *278*, 62–78. [CrossRef]
14. Li, Z.; Das, S.; Hongmanorom, P.; Dewangan, N.; Wai, M.H.; Kawi, S. Silica-based micro- and mesoporous catalysts for dry reforming of methane. *Catal. Sci. Technol.* **2018**, *8*, 2763–2778. [CrossRef]
15. Bian, Z.; Kawi, S. Preparation, characterization and catalytic application of phyllosilicate: A review. *Catal. Today* **2020**, *339*, 3–23. [CrossRef]
16. Gao, X.; Ashok, J.; Kawi, S. Smart Designs of Anti-Coking and Anti-Sintering Ni-Based Catalysts for Dry Reforming of Methane: A Recent Review. *Reactions* **2020**, *1*, 162–194. [CrossRef]
17. Jang, W.-J.; Shim, J.-O.; Kim, H.-M.; Yoo, S.-Y.; Roh, H.-S. A review on dry reforming of methane in aspect of catalytic properties. *Catal. Today* **2019**, *324*, 15–26. [CrossRef]
18. Li, K.; Liu, J.-L.; Li, X.-S.; Lian, H.-Y.; Zhu, X.; Bogaerts, A.; Zhu, A.-M. Novel power-to-syngas concept for plasma catalytic reforming coupled with water electrolysis. *Chem. Eng. J.* **2018**, *353*, 297–304. [CrossRef]
19. Uytdenhouwen, Y.; Bal, K.; Neyts, E.; Meynen, V.; Cool, P.; Bogaerts, A. On the kinetics and equilibria of plasma-based dry reforming of methane. *Chem. Eng. J.* **2021**, *405*, 6630. [CrossRef]
20. Ray, D.; Reddy, P.K.R.; Subrahmanyam, C. Ni-Mn/γ-Al2O3 assisted plasma dry reforming of methane. *Catal. Today* **2018**, *309*, 212–218. [CrossRef]
21. Brune, L.; Ozkan, A.; Genty, E.; De Bocarmé, T.V.; Reniers, F. Dry reforming of methane via plasma-catalysis: Influence of the catalyst nature supported on alumina in a packed-bed DBD configuration. *J. Phys. D Appl. Phys.* **2018**, *51*, 4002. [CrossRef]
22. Rico, V.J.; Hueso, J.L.; Cotrino, J.; Gonzalez-Elipe, A.R. Evaluation of different dielectric barrier discharge plasma configura-tions as an alternative technology for green C1 chemistry in the carbon dioxide reforming of methane and the direct decom-position of methanol. *J. Phys. Chem. A* **2010**, *114*, 4009–4016. [CrossRef]

23. Vakili, R.; Gholami, R.; Stere, C.E.; Chansai, S.; Chen, H.; Holmes, S.M.; Jiao, Y.; Hardacre, C.; Fan, X. Plasma-assisted catalytic dry reforming of methane (DRM) over metal-organic frameworks (MOFs)-based catalysts. *Appl. Catal. B Environ.* **2020**, *260*, 8195. [CrossRef]

24. Abiev, R.S.; Sladkovskiy, D.A.; Semikin, K.V.; Murzin, D.Y.; Rebrov, E.V. Non-Thermal Plasma for Process and Energy Intensification in Dry Reforming of Methane. *Catalysts* **2020**, *10*, 1358. [CrossRef]

25. Li, Z.; Lin, Q.; Li, M.; Cao, J.; Liu, F.; Pan, H.; Wang, Z.; Kawi, S. Recent advances in process and catalyst for CO2 reforming of methane. *Renew. Sustain. Energy Rev.* **2020**, *134*, 312. [CrossRef]

26. George, A.; Shen, B.; Craven, M.; Wang, Y.; Kang, D.; Wu, C.; Tu, X. A Review of Non-Thermal Plasma Technology: A novel solution for CO2 conversion and utilization. *Renew. Sustain. Energy Rev.* **2021**, *135*, 9702. [CrossRef]

27. Zeng, Y.; Zhu, X.; Mei, D.; Ashford, B.; Tu, X. Plasma-catalytic dry reforming of methane over γ-Al2O3 supported metal catalysts. *Catal. Today* **2015**, *256*, 80–87. [CrossRef]

28. Zhang, J.; Wang, H.; Dalai, A.K. Effects of metal content on activity and stability of Ni-Co bimetallic catalysts for CO2 reforming of CH. *Appl. Catal. A Gen.* **2008**, *339*, 121–129. [CrossRef]

29. Theofanidis, S.A.; Galvita, V.V.; Poelman, H.; Marin, G.B. Enhanced Carbon-Resistant Dry Reforming Fe-Ni Catalyst: Role of Fe. *ACS Catal.* **2015**, *5*, 3028–3039. [CrossRef]

30. Fan, M.S.; Abdullah, A.Z.; Bhatia, S. Kinetic Evaluation of Methane—Carbon Dioxide Reforming Process Based on the Reaction Steps. *Chem. Sus. Chem.* **2011**, *4*, 1643–1653. [CrossRef]

31. Li, Z.; Kathiraser, Y.; Ashok, J.; Oemar, U.; Kawi, S. Simultaneous Tuning Porosity and Basicity of Nickel@Nickel–Magnesium Phyllosilicate Core–Shell Catalysts for CO2 Reforming of CH. *Langmuir* **2014**, *30*, 14694–14705. [CrossRef] [PubMed]

32. Gao, X.; Tan, Z.; Hidajat, K.; Kawi, S. Highly reactive Ni-Co/SiO2 bimetallic catalyst via complexation with oleylamine/oleic acid organic pair for dry reforming of methane. *Catal. Today* **2017**, *281*, 250–258. [CrossRef]

33. Oemar, U.; Kathiraser, Y.; Mo, L.; Ho, X.K.; Kawi, S. CO2reforming of methane over highly active La-promoted Ni supported on SBA-15 catalysts: Mechanism and kinetic modelling. *Catal. Sci. Technol.* **2016**, *6*, 1173–1186. [CrossRef]

34. Ni, J.; Zhao, J.; Chen, L.; Lin, J.; Kawi, S. Lewis Acid Sites Stabilized Nickel Catalysts for Dry (CO2) Reforming of Methane. *ChemCatChem* **2016**, *8*, 3732–3739. [CrossRef]

35. Mo, L.; Leong, K.K.M.; Kawi, S. A highly dispersed and anti-coking Ni-La2O3/SiO2 catalyst for syngas production from dry carbon dioxide reforming of methane. *Catal. Sci. Technol.* **2014**, *4*, 2107–2114. [CrossRef]

36. Ni, J.; Chen, L.; Lin, J.; Schreyer, M.K.; Wang, Z.; Kawi, S. High performance of Mg–La mixed oxides supported Ni catalysts for dry reforming of methane: The effect of crystal structure. *Int. J. Hydrog. Energy* **2013**, *38*, 13631–13642. [CrossRef]

37. Tu, X.; Whitehead, J.C. Plasma-catalytic dry reforming of methane in an atmospheric dielectric barrier discharge: Under-standing the synergistic effect at low temperature. *Appl. Catal. B Environ.* **2012**, *125*, 439–448. [CrossRef]

38. Mei, D.; Ashford, B.; He, Y.-L.; Tu, X. Plasma-catalytic reforming of biogas oversupported Ni catalysts in a dielectric barrier discharge reactor: Effect of catalyst supports. *Plasma Process. Polym.* **2017**, *14*, 1600076. [CrossRef]

39. Bogaerts, A.; Centi, G. Plasma Technology for CO2 Conversion: A Personal Perspective on Prospects and Gaps. *Front. Energy Res.* **2020**, *8*, 111. [CrossRef]

40. Jennifer, M.-C.; Coulombe, S.; Kopyscinski, J. Influence of Operating Parameters on Plasma-Assisted Dry Reforming of Methane in a Rotating Gliding Arc Reactor. *Plasma Chem. Plasma Process.* **2020**, *40*, 857–881.

41. Kogelschatz, U. Dielectric-Barrier Discharges: Their History, Discharge Physics, and Industrial Applications. *Plasma Chem. Plasma Process.* **2003**, *23*, 1–46. [CrossRef]

42. Zhang, X.; Wenren, Y.; Zhou, W.; Han, J.; Lu, H.; Zhu, Z.; Wu, Z.; Cha, M.S. Dry reforming of methane in a temperature-controlled dielectric barrier discharge reactor: Disclosure of reactant effect. *J. Phys. D Appl. Phys.* **2020**, *53*, 4002. [CrossRef]

43. Khoja, A.H.; Tahir, M.; Amin, N.A.S. Cold plasma dielectric barrier discharge reactor for dry reforming of methane over Ni/γ-Al2O3-MgO nanocomposite. *Fuel Process. Technol.* **2018**, *178*, 166–179. [CrossRef]

44. Andersen, J.; Christensen, J.; Østberg, M.; Bogaerts, A.; Jensen, A. Plasma-catalytic dry reforming of methane: Screening of catalytic materials in a coaxial packed-bed DBD reactor. *Chem. Eng. J.* **2020**, *397*, 5519. [CrossRef]

45. Michielsen, I.; Uytdenhouwen, Y.; Bogaerts, A.; Meynen, V. Altering Conversion and Product Selectivity of Dry Reforming of Methane in a Dielectric Barrier Discharge by Changing the Dielectric Packing Material. *Catalysts* **2019**, *9*, 51. [CrossRef]

46. Chun, S.M.; Shin, D.H.; Ma, S.H.; Yang, G.W.; Hong, Y.C. CO2 Microwave Plasma—Catalytic Reactor for Efficient Reforming of Methane to Syngas. *Catalysts* **2019**, *9*, 292. [CrossRef]

47. Holzer, F.; Kopinke, F.D.; Roland, U. Influence of Ferroelectric Materials and Catalysts on the Performance of Non-Thermal Plasma (NTP) for the Removal of Air Pollutants. *Plasma Chem. Plasma Process.* **2005**, *25*, 595–611. [CrossRef]

48. Li, H.; Yuan, G.; Shan, B.; Zhang, X.; Ma, H.; Tian, Y.; Lu, H.; Liu, J. Chemical Vapor Deposition of Vertically Aligned Carbon Nanotube Arrays: Critical Effects of Oxide Buffer Layers. *Nanoscale Res. Lett.* **2019**, *14*, 106. [CrossRef] [PubMed]

49. Kameshima, S.; Tamura, K.; Ishibashi, Y.; Nozaki, T. Pulsed dry methane reforming in plasma-enhanced catalytic reaction. *Catal. Today* **2015**, *256*, 67–75. [CrossRef]

50. Slaets, J.; Aghaei, M.; Ceulemans, S.; Van Alphen, S.; Bogaerts, A. CO2 and CH4 conversion in "real" gas mixtures in a gliding arc plasmatron: How do N2 and O2 affect the performance? *Green Chem.* **2020**, *22*, 1366–1377. [CrossRef]

51. Damjan, L.J.; Lin, J.-L.; Pohar, A.; Likozar, B. Methane Dry Reforming over Ni/Al2O3 Catalyst in Spark Plasma Reactor: Linking Computational Fluid Dynamics (CFD) with Reaction Kinetic Modelling. *Catal. Today* **2021**, *362*, 11–12.

52. Yap, D.; Tatibouët, J.-C.; Catherine, B.-D. Catalyst assisted by non-thermal plasma in dry reforming of methane at low tem-perature. *Catal. Today* **2018**, *299*, 263–271. [CrossRef]
53. Ray, D.; Nepak, D.; Vinodkumar, T.; Subrahmanyam, C. g-C3N4 promoted DBD plasma assisted dry reforming of methane. *Energy* **2019**, *183*, 630–638. [CrossRef]
54. Khoja, A.H.; Tahir, M.; Amin, N.A.S. Evaluating the Performance of a Ni Catalyst Supported on La2O3-MgAl2O4 for Dry Reforming of Methane in a Packed Bed Dielectric Barrier Discharge Plasma Reactor. *Energy Fuels* **2019**, *33*, 11630–11647. [CrossRef]
55. Hu, X.; Jia, X.; Zhang, X.; Liu, Y.; Liu, C.-J. Improvement in the activity of Ni/ZrO2 by cold plasma decomposition for dry reforming of methane. *Catal. Commun.* **2019**, *128*, 5720. [CrossRef]
56. Sun, Y.; Zhang, G.; Liu, J.; Zhao, P.; Hou, P.; Xu, Y.; Zhang, R. Effect of different activated carbon support on CH4-CO2 re-forming over Co-based catalysts. *Int. J. Hydr. Energy* **2018**, *43*, 1497–1507. [CrossRef]
57. Effendi, A.; Hellgardt, K.; Zhang, Z.-G.; Yoshida, T. Characterisation of carbon deposits on Ni/SiO2 in the reforming of CH4–CO2 using fixed- and fluidised-bed reactors. *Catal. Commun.* **2003**, *4*, 203–207. [CrossRef]
58. Luo, Y.-R. *Comprehensive Handbook of Chemical Bond Energies*; Chapman and Hall/CRC: London, UK; Boca Raton, FL, USA, 2007.
59. Zhang, K.; Mukhriza, T.; Liu, X.; Greco, P.P.; Chiremba, E. A study on CO2 and CH4 conversion to synthesis gas and higher hydrocarbons by the combination of catalysts and dielectric-barrier discharges. *Appl. Catal. A Gen.* **2015**, *502*, 138–149. [CrossRef]
60. Wang, H.; Han, J.; Bo, Z.; Qin, L.; Wang, Y.; Yu, F. Non-thermal plasma enhanced dry reforming of CH4 with CO2 over activated carbon supported Ni catalysts. *Mol. Catal.* **2019**, *475*, 486. [CrossRef]
61. Zeng, Y.; Wang, L.; Wu, C.; Wang, J.; Shen, B.; Tu, X. Low temperature reforming of biogas over K-, Mg- and Ce-promoted Ni/Al2O3 catalysts for the production of hydrogen rich syngas: Understanding the plasma-catalytic synergy. *Appl. Catal. B Environ.* **2018**, *224*, 469–478. [CrossRef]
62. Ray, D.; Nepak, D.; Janampelli, S.; Goshal, P.; Subrahmanyam, C.; Ghosal, P.; Subrahmanyam, C. Dry Reforming of Methane in DBD Plasma over Ni-Based Catalysts: Influence of Process Conditions and Support on Performance and Durability. *Energy Technol.* **2019**, *7*, 1008. [CrossRef]
63. Suttikul, T.; Nuchdang, S.; Rattanaphra, D.; Phalakornkule, C. Influence of Operating Parameters, Al2O3 and Ni/Al2O3 Cata-lysts on Plasma Assisted CO2 Reforming of CH4 in a Parallel Plate Dielectric Barrier Discharge for High H2/CO Ratio Syngas Production. *Plasma Chem. Plasma Process.* **2020**, *40*, 1445–1463. [CrossRef]
64. Kameshima, S.; Mizukami, R.; Yamazaki, T.; Prananto, L.A.; Nozaki, T. Interfacial reactions between DBD and porous catalyst in dry methane reforming. *J. Phys. D: Appl. Phys.* **2018**, *51*, 4006. [CrossRef]
65. Nagaraja, B.M.; Bulushev, D.A.; Beloshapkin, S.; Ross, J.R.H. The effect of potassium on the activity and stability of Ni-MgO-ZrO2 catalysts for the dry reforming of methane to give synthesis gas. *Catal. Today* **2011**, *178*, 132–136. [CrossRef]
66. Vasiliades, M.A.; Djinović, P.; Pintar, A.; Kovač, J.; Efstathiou, A.M. The effect of CeO2–ZrO2 structural differences on the origin and reactivity of carbon formed during methane dry reforming over NiCo/CeO2–ZrO2 catalysts studied by transient techniques. *Catal. Sci. Technol.* **2017**, *7*, 5422–5434. [CrossRef]
67. Hou, Z.; Yashima, T. Small Amounts of Rh-Promoted Ni Catalysts for Methane Reforming with CO. *Catal. Lett.* **2003**, *89*, 193–197. [CrossRef]
68. Rahemi, N.; Haghighi, M.; Babaluo, A.A.; Allahyari, S.; Estifaee, P.; Jafari, M.F. Plasma-Assisted Dispersion of Bimetallic Ni–Co over Al2O3–ZrO2 for CO2 Reforming of Methane: Influence of Voltage on Catalytic Properties. *Top. Catal.* **2017**, *60*, 145–854. [CrossRef]
69. Mei, D.; Liu, S.; Tu, X. CO2 reforming with methane for syngas production using a dielectric barrier discharge plasma coupled with Ni/γ-Al2O3 catalysts: Process optimization through response surface methodology. *J. CO2 Util.* **2017**, *21*, 314–326. [CrossRef]
70. Dębek, R.; Motak, M.; Galvez, M.E.; Da Costa, P.; Grzybek, T. Catalytic activity of hydrotalcite-derived catalysts in the dry reforming of methane: On the effect of Ce promotion and feed gas composition. *React. Kinet. Mech. Catal.* **2017**, *121*, 185–208. [CrossRef]
71. Yan, X.; Hu, T.; Liu, P.; Li, S.; Zhao, B.; Zhang, Q.; Jiao, W.; Chen, S.; Wang, P.; Lu, J.; et al. Highly efficient and stable Ni/CeO2-SiO2 catalyst for dry reforming of methane: Effect of interfacial structure of Ni/CeO2 on SiO. *Appl. Catal. B: Environ.* **2019**, *246*, 221–231. [CrossRef]
72. Jin, L.; Li, Y.; Lin, P.; Hu, H. CO2 reforming of methane on Ni/γ-Al2O3 catalyst prepared by dielectric barrier discharge hydrogen plasm. *Int. J. Hydrog. Energy* **2014**, *39*, 5756–5763. [CrossRef]
73. Chen, H.-W.; Wang, C.-Y.; Yu, C.-H.; Tseng, L.-T.; Liao, P.-H. Carbon dioxide reforming of methane reaction catalyzed by stable nickel copper catalysts. *Catal. Today* **2004**, *97*, 173–180. [CrossRef]
74. Dębek, R.; Motak, M.; Grzybek, T.; Galvez, M.E.; Da Costa, P. A Short Review on the Catalytic Activity of Hydrotalcite-Derived Materials for Dry Reforming of Methane. *Catalysts* **2017**, *7*, 32. [CrossRef]
75. Tao, X.; Yang, C.; Huang, L.; Xu, D. DBD plasma combined with catalysts derived from NiMgAlCe hydrotalcite for CO2 reforming of CH. *Mater. Chem. Phys.* **2020**, *250*, 3118. [CrossRef]
76. Sengupta, S.; Deo, G. Modifying alumina with CaO or MgO in supported Ni and Ni–Co catalysts and its effect on dry reforming of CH. *J. CO2 Util.* **2015**, *10*, 67–77. [CrossRef]
77. Özkara-Aydınoğlu, Ş.; Aksoylu, A.E. Carbon dioxide reforming of methane over Co-X/ZrO2 catalysts (X = La, Ce, Mn, Mg, K). *Catal. Commun.* **2010**, *11*, 1165–1170. [CrossRef]

78. Gurav, H.R.; Dama, S.; Samuel, V.; Chilukuri, S. Influence of preparation method on activity and stability of Ni catalysts supported on Gd doped ceria in dry reforming of methane. *J. CO2 Util.* **2017**, *20*, 357–367. [CrossRef]
79. Bacariza, M.; Biset-Peiró, M.; Graça, I.; Guilera, J.; Morante, J.; Lopes, J.; Andreu, T.; Henriques, C. DBD plasma-assisted CO2 methanation using zeolite-based catalysts: Structure composition-reactivity approach and effect of Ce as promoter. *J. CO2 Util.* **2018**, *26*, 202–211. [CrossRef]
80. Ray, D.; Chawdhury, P.; Subrahmanyam, C. Promising Utilization of CO2 for Syngas Production over Mg2+- and Ce2+- Pro-moted Ni/γ-Al2O3 Assisted by Nonthermal Plasma. *ACS Omega* **2020**, *5*, 14040–14050. [CrossRef]
81. Sheng, Z.; Kim, H.-H.; Yao, S.; Nozaki, T. Plasma-chemical promotion of catalysis for CH4 dry reforming: Unveiling plas-ma-enabled reaction mechanisms. *Phys. Chem. Chem. Phys.* **2020**, *22*, 19349–19358. [CrossRef]
82. Zheng, X.; Tan, S.; Dong, L.; Li, S.; Chen, H. LaNiO3@SiO2 core–shell nano-particles for the dry reforming of CH4 in the dielectric barrier discharge plasma. *Int. J. Hydrog. Energy* **2014**, *39*, 11360–11367. [CrossRef]
83. Zheng, X.-G.; Tan, S.-Y.; Dong, L.-C.; Li, S.-B.; Chen, H.-M.; Wei, S.-A. Experimental and kinetic investigation of the plasma catalytic dry reforming of methane over perovskite LaNiO3 nanoparticles. *Fuel Process. Technol.* **2015**, *137*, 250–258. [CrossRef]
84. Choi, J.H.; Il, L.T.; Han, I.; Oh, B.Y.; Jeong, M.C.; Myoung, J.M. Improvement of plasma uniformity using ZnO-coated dielectric barrier discharge in open air. *Appl. Phys. Lett.* **2006**, *89*, 8150. [CrossRef]
85. Holzer, F.; Roland, U.; Kopinke, F.D. Combination of non-thermal plasma and heterogeneous catalysis for oxidation of volatile organic compounds: Part 1. accessibility of the intra-particle volume. *Appl. Catal. B.* **2002**, *38*, 163–181. [CrossRef]
86. Ozkan, A.; Dufour, T.; Bogaerts, A.; Reniers, F. How do the barrier thickness and dielectric material influence the filamentary mode and CO2 conversion in a flowing DBD? *Plasma Sources Sci. Technol.* **2016**, *25*, 5016. [CrossRef]
87. Khoja, A.H.; Tahir, M.; Amin, N.A.S. Dry reforming of methane using different dielectric materials and DBD plasma reactor configurations. *Energy Convers. Manag.* **2017**, *144*, 262–274. [CrossRef]
88. Kundu, S.K.; Kennedy, E.M.; Gaikwad, V.V.; Molloy, T.S.; Dlugogorski, B.Z. Experimental investigation of alumina and quartz as dielectrics for a cylindrical double dielectric barrier discharge reactor in argon diluted methane plasma. *Chem. Eng. J.* **2012**, *180*, 178–189. [CrossRef]
89. Yan, X.; Zhao, B.; Liu, Y.; Li, Y. Dielectric barrier discharge plasma for preparation of Ni-based catalysts with enhanced coke resistance: Current status and perspective. *Catal. Today* **2015**, *256*, 29–40. [CrossRef]
90. Duan, X.; Li, Y.; Ge, W.; Wang, B. Degradation of CO2through dielectric barrier discharge microplasma. *Greenh. Gases: Sci. Technol.* **2015**, *5*, 131–140. [CrossRef]
91. Pinhão, N.; Moura, A.; Branco, J.; Neves, J. Influence of gas expansion on process parameters in non-thermal plasma plug-flow reactors: A study applied to dry reforming of methane. *Int. J. Hydrog. Energy* **2016**, *41*, 9245–9255. [CrossRef]
92. Chawdhury, P.; Ray, D.; Subrahmanyam, C. Single step conversion of methane to methanol assisted by nonthermal plasma. *Fuel Process. Technol.* **2018**, *179*, 32–41. [CrossRef]
93. Mao, S.; Tan, Z.; Zhang, L.; Huang, Q. Plasma-assisted biogas reforming to syngas at room temperature condition. *J. Energy Inst.* **2018**, *91*, 172–183. [CrossRef]
94. Liu, C.; Li, M.; Wang, J.; Zhou, X.; Guo, Q.; Yan, J.; Li, Y. Plasma methods for preparing green catalysts: Current status and perspective. *Chin. J. Catal.* **2016**, *37*, 340–348. [CrossRef]
95. Michielsen, I.; Uytdenhouwen, Y.; Pype, J.; Mertens, J.; Reniers, F.; Meynen, V.; Bogaerts, A. CO2 dissociation in a packed bed DBD reactor: First steps towards a better understanding of plasma catalysis. *Chem. Eng. J.* **2017**, *326*, 477–488. [CrossRef]
96. Zhang, X.; Cha, M.S. Electron-induced dry reforming of methane in a tempera-ture-controlled dielectric barrier discharge reactor. *J. Phys. D Appl. Phys.* **2013**, *46*, 5205. [CrossRef]
97. Snoeckx, R.; Bogaerts, A. Plasma technology–A novel solution for CO2 conversion? *Chem. Soc. Rev.* **2017**, *46*, 5805–5863. [CrossRef] [PubMed]
98. Sheng, Z.; Kameshima, S.; Sakata, K.; Nozaki, T. Plasma-Enabled Dry Methane Reforming. *Plasma Chem. Gas Convers.* **2018**. [CrossRef]

Article

The Potential Use of Core-Shell Structured Spheres in a Packed-Bed DBD Plasma Reactor for CO$_2$ Conversion

Yannick Uytdenhouwen [1,2], Vera Meynen [2], Pegie Cool [2] and Annemie Bogaerts [1,*]

[1] Research Group PLASMANT, Department of Chemistry, University of Antwerp, Universiteitsplein 1, B-2610 Wilrijk, Belgium; yannick.uytdenhouwen@uantwerpen.be
[2] Research Group LADCA, Department of Chemistry, University of Antwerp, Universiteitsplein 1, B-2610 Wilrijk, Belgium; vera.meynen@uantwerpen.be (V.M.); pegie.cool@uantwerpen.be (P.C.)
* Correspondence: annemie.bogaerts@uantwerpen.be

Received: 2 April 2020; Accepted: 9 May 2020; Published: 11 May 2020

Abstract: This work proposes to use core-shell structured spheres to evaluate whether it allows to individually optimize bulk and surface effects of a packing material, in order to optimize conversion and energy efficiency. Different core-shell materials have been prepared by spray coating, using dense spheres (as core) and powders (as shell) of SiO$_2$, Al$_2$O$_3$, and BaTiO$_3$. The materials are investigated for their performance in CO$_2$ dissociation and compared against a benchmark consisting of a packed-bed reactor with the pure dense spheres, as well as an empty reactor. The results in terms of CO$_2$ conversion and energy efficiency show various interactions between the core and shell material, depending on their combination. Al$_2$O$_3$ was found as the best core material under the applied conditions here, followed by BaTiO$_3$ and SiO$_2$, in agreement with their behaviour for the pure spheres. Applying a thin shell layer on the cores showed equal performance between the different shell materials. Increasing the layer thickness shifts this behaviour, and strong combination effects were observed depending on the specific material. Therefore, this method of core-shell spheres has the potential to allow tuning of the packing properties more closely to the application by designing an optimal combination of core and shell.

Keywords: plasma; plasma catalysis; dielectric barrier discharge; CO$_2$ dissociation; core-shell spheres; packed-bed reactor

1. Introduction

A dielectric barrier discharge (DBD) reactor is a popular type of plasma reactor used for a variety of reactions, including both decomposition and synthesis reactions [1–4]. It is flexible in use, i.e., power, frequency, pressure, discharge gap, reactor shape, and flow patterns can be varied, and it can also be easily upscaled and implemented for industrial use [3]. The performance of a DBD reactor can be improved by adding a packing material to the reaction zone to obtain higher conversions, selectivities, and/or energy efficiencies compared to the standard empty DBD reactor [1,2,5–7]. Adding a packing material to the reaction zone will induce both physical and chemical changes, resulting in a wide variety of outcomes [4,8]. On one hand, the packing material will change physical aspects such as the gas flow behaviour through the reactor–by reducing the discharge volume to small voids between the spheres, altering the flow and mixing patterns, and reducing the residence time–as well as the characteristics of the plasma and the discharging behaviour. The properties of the packing material, i.e., size and shape, dielectric constant, (elemental) composition, surface roughness, thermal and electrical properties, porosity, etc., can influence the type of discharge, electron temperature and density, surface losses, etc. [8–10]. We can differentiate the effects of material properties into bulk and surface effects, e.g., dielectric constant and electrical conductivity are bulk effects, while surface roughness and adsorption are surface effects. On the other hand, the reactive plasma can influence the packing materials as well, both on a short and long term. Short term effects include the generation of a direct flux of excited species, radicals, or ions towards the surface, lowering the activation barrier, and changing reaction pathways; long term effects are alterations to the material structure, such as changing oxidation states, etching/destruction of the surface/pores, and/or inactivation of doped catalysts [8]. Furthermore, the gas mixture itself will influence the plasma characteristics, and requires specific conditions as well for optimal transfer of electrical energy in chemical energy. It is therefore not at all straightforward to correlate any cause and effect, when introducing and comparing different packing materials.

When searching for the best packing material, using pure, dense bulk materials quickly hits some obstacles, as each material has its fixed and specific properties that cannot be changed individually at the surface and in the bulk. Moreover, changing the material type varies several of the above-mentioned parameters (both physical and chemical; and both surface and bulk), that impact the plasma behaviour and surface chemistry at the same time. Indeed, in our previous work we investigated millimetre-sized spheres from different materials in a packed-bed DBD reactor for both CO_2 dissociation (acting as benchmark in this work as well) and dry reforming of methane (DRM), and we found it was not a straightforward method to pinpoint the exact origin of the observed effects [5,6,11]. Additionally, using exactly the same reactor with different operating parameters renders different results [5,6]. Hence, in order to be able to better tune a packing material, combining different properties in bulk (physical) and surface (physical-chemical) behaviour might be needed for optimal performance. The combination of these properties might, however, not be present in one type of material. Therefore, we evaluate in this paper the potential of using millimetre-sized core-shell structured spheres. These spheres consist, as the name implies, of a dense spherical core that is covered with a (thin) shell (of 50 to 500 μm thickness). The core will mainly determine the bulk effect of the entire sphere such as dielectric constant, thermal and electrical properties; while the shell, will determine mainly the surface effects such as porosity, adsorption, (electrical) surface properties, and surface chemistry, as well as some bulk effects, if the shell is made sufficiently thick—potentially shielding core effects. With this approach, we can test a wide variety of combinations of core and shell material types, and their respective sizes, in order to evaluate the potential of core-shell materials to tune the DBD reactor performance.

Core-shell spheres have been widely used in the past, with examples found in thermal catalysis, electrocatalysis, photocatalysis, drug screening, etc. [12–20], and with coated pellets widely used in the pharmaceutical and food industry [21–23]. Usually the core-shell spheres are produced in the micro- to nanometre range via methods such as sol-gel deposition, hydrothermal synthesis, colloidal synthesis, plasma deposition, and microfluidic gelation; with only a few examples found in the millimetre-size range made by hydrothermal synthesis or spray coating [13,21–24]. Although being widely used in different fields of research and application, core-shell spheres are rarely adopted in plasma conversion processes, with only a few cases reported in literature, e.g., Zheng et al. used nano-sized core-shell particles for DRM [25,26]. Coated spheres with dispersed or clustered catalytically active materials, or nano-sized (mono) layers, have also been used in plasma reactors [27], but to our knowledge, no research has been reported using millimetre-sized core-shell spheres with systematically altered core-shell combinations of different materials as those we propose here.

This approach is evaluated for CO_2 splitting into CO and O_2 due to its simpler chemistry compared to multi-component mixtures, such as (dry) reforming of methane. Specifically, we will investigate the influence of adding a shell to a spherical core, and how the respective material combinations and shell thickness affect the overall performance of the DBD reactor, in terms of CO_2 conversion and energy efficiency, in order to estimate its potential in design of appropriate packing materials for plasma conversion processes. The purpose is not finding the highest activity but evaluating the potential of core-shell structures to improve packed-bed plasma reactor performance.

2. Results

Before discussing the effect of using core-shell spheres with different shell thicknesses, the reference will be discussed (empty reactor and pure cores). Note that the results of the pure spheres were obtained in our previous work and are more thoroughly discussed there [5].

2.1. Benchmark Measurements for the Empty Reactor and the Reactor Packed with Pure Spheres

Figure 1a shows the results for the empty reactor at the standard conditions (denoted as "=Flow"), and a higher flow rate to obtain the same residence time (RT) as the packed reactors (denoted as "=RT"). It shows a base conversion and energy efficiency of 11% and 3.0%, respectively, for a flow rate of 38.98 mL/min (27.20 s RT); while obtaining 6.4% and 3.2%, respectively, for a flow rate of 79.96 mL/min (14.07 s RT). Note that all but one bar in Figure 1 represent both conversion and energy efficiency since they are just scaled. The measurement of the empty reactor at the same residence time ("=RT") has two parts of which the lower part depicts the conversion and the upper part the energy efficiency.

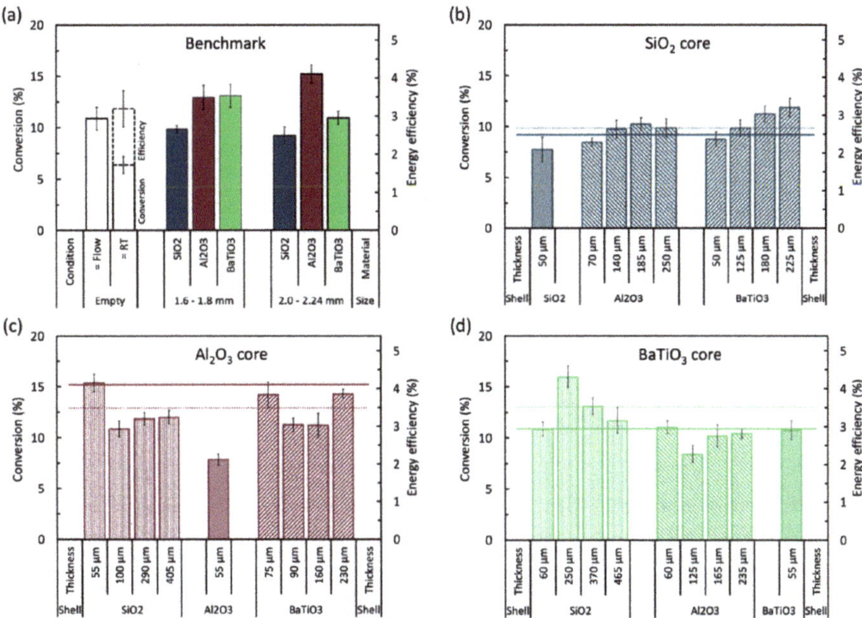

Figure 1. Overview graph of the conversion and energy efficiency of (**a**) the benchmark results (i.e., empty reactor and packed reactor with pure spheres [5]), and all core-shell samples with (**b**) SiO_2 core, (**c**) Al_2O_3 core, and (**d**) $BaTiO_3$ core. All core-shell samples were coated with SiO_2, Al_2O_3, and $BaTiO_3$ powder in different shell thicknesses, as indicated in each figure. All measurements were performed at 30W, 38.98 mL/min, and 1 bar; except for the empty reactor, which was also evaluated at the same residence time as the packed reactors (79.96 mL/min). The dotted and solid lines are the performance of the corresponding core spheres in their small and big size, respectively. The exact values can be found in Table A1 in Appendix A.

Inserting a packing material into the discharge gap shows clear material and size dependent effects, as also shown and discussed in [5,6]. All packed reactors show better conversions than the empty reactor at the same residence time, i.e., same gas treatment time. However, when compared with the empty reactor at the same flow rate, i.e., same throughput, only the Al_2O_3 spheres, and the $BaTiO_3$ spheres at 1.6 to 1.8 mm diameter perform better (both in conversion and energy efficiency). SiO_2 with a size range of 1.6 to 1.8 mm shows a lower conversion of 9.8% and the larger sphere size, 2.0 to 2.24

mm, even yields a slightly worse conversion of 9.2% (within error bars). This means that SiO_2 is able to enhance the electric properties of the plasma through local electric field enhancements [28], but not enough to compensate for the 48.27% reduction in reaction volume due to the spheres (see Section 4.2). Adding a packing to the reactor increases the available surface area to promote (catalytic) surface chemistry. If present, this surface chemistry can also inhibit the plasma chemistry, by losses of energetic species upon collision with the surface [29]. Al_2O_3, however, can compensate for the reaction volume reduction, with a conversion of 13% at a 1.6 to 1.8 mm sphere size, and 15.2% at a sphere size of 2.0 to 2.24 mm. Lastly, $BaTiO_3$ shows an improved conversion of 13% at 1.6 to 1.8 mm sphere size, but the 2.0 to 2.24 mm spheres performed worse, with a conversion of 10.9%. These results illustrate the interaction between sphere size and material type on the conversion. Both SiO_2 and $BaTiO_3$ exhibit a slight decrease in conversion, while Al_2O_3 shows an increase at larger sphere size. This is consistent with the modelling work of Van Laer and Bogaerts [9], which revealed that there is not necessarily a linear correlation between dielectric constant and the plasma parameters (electric field, electron temperature, and electron density), as well as the sphere size (number of spheres in the discharge gap). The trends in energy efficiency are the same as for the conversion, which is logical when the flow rate is kept constant. More considerations about the energy efficiency will be given in Section 3 below.

Evidently, each material behaves differently based on their size, and this may be attributed to a number of material specific properties–such as dielectric constant, surface roughness, surface chemistry, electrical conductivity, heat capacity, etc., as well as the number of contact points, void space between the spheres, etc., which proved to be difficult to correlate by previous researchers [6,11]. By comparing these benchmark results with the performance of the core-shell samples in the next section, we hope to obtain some clues on the effect of the material properties with respect to their bulk or surface effect.

2.2. Core-Shell Spheres

A matrix of samples has been prepared based on three materials (SiO_2, Al_2O_3, and $BaTiO_3$) in different shell thicknesses to investigate the effect of the shell material, the core material, and the shell thickness, as shown in Figure 1b–d. However, for an unbiased evaluation, first we will look into the actual influence of adding a shell to the spheres, by coating them with a shell of the same material as the core material.

2.2.1. Influence of a Shell on the Performance of the Spheres

By coating the pure spheres with a thin layer of the same powderous material as the core, we can investigate how a calcined powder layer affects the performance compared to the pure spheres. Figure 2 shows the results of the pure spheres coated with a thin layer of approximately 50 μm to form a shell of the same material. As can be observed, all core-shell spheres show worse results than their pure non-coated 1.6 to 1.8 mm variants (shown as a solid line). $SiO_2@SiO_2$ has the smallest drop in conversion (i.e., 8% compared to 9.8% for the pure sphere), $BaTiO_3@BaTiO_3$ shows a somewhat larger drop in conversion (i.e., 10.8% compared to 13%), but the biggest difference is seen with $Al_2O_3@Al_2O_3$ where the conversion drops to 7.8% compared to 12.8% for the pure spheres. This negative effect of adding a shell of the same material may be attributed to a different morphology (i.e., macroscopic roughness, powder grain surface, introduced interstitial porosity between the grains, etc.), a negative effect of the LUDOX binder, or a masking effect by the shell for a useful property of the core, or a combination of all effects. Indeed, it has been shown that the morphology of the spheres can have a significant impact on the chemistry in plasmas [8]. This is because the morphology can enhance the local electric field by extra contact points, sharp edges, and close surfaces in for example macropores, resulting in different plasma discharges, changing the local chemistry. This may or may not have an effect on the performance, which can be either beneficial or detrimental, depending on the reactions. Additionally, the added LUDOX binder, although necessary for shaping, is an extra material added to the shell which might introduce an unknown effect on the shell performance. Binder effects are known for thermal catalysis [30–32], but have not yet been studied in plasma catalysis, as far as we know. It is also for this reason that we kept the amount of binder limited to 1 wt%, although we cannot exclude its effect even at these small quantities. Furthermore, as plasma can only be formed in pores with diameter larger than the Debye length, which is typically several 100 nm for DBD plasma conditions [33], most of the bulk material of the spheres might be out of reach for the reactive plasma species to promote any reaction. This means that adding a shell, although porous in nature, can shield (most of) the reactive plasma species from the core and thus inhibit any activated core surface chemistry. Apparently, this effect is more present with the $Al_2O_3@Al_2O_3$ spheres. Exactly the same source powder of α-Al_2O_3 was used to prepare the core spheres and to coat the shells, but they were subject to different synthesis processes (e.g., thermal post-treatments) and different chemicals (e.g., calcium present in the core, a silica binder in the shell). This hints that either (i) the pure Al_2O_3 spheres exhibit an effect that is particularly well masked by the shell, e.g., the presence of calcium compounds remaining from the synthesis ($CaCO_3$, CaO, and/or $CaO.xAl_2O_3$ with x = 1, 2, or 6) [34]; and/or (ii) the Al_2O_3 powder has a large inhibiting effect as shell material, by lacking desired surface properties, having unwanted porosity, the presence of the LUDOX binder, or another, yet unidentified, aspect. This clearly indicates that a particular difference in the shell can induce physical aspects that have a relatively large impact on conversion and energy efficiency, necessitating further studies with samples that are well controlled in these properties.

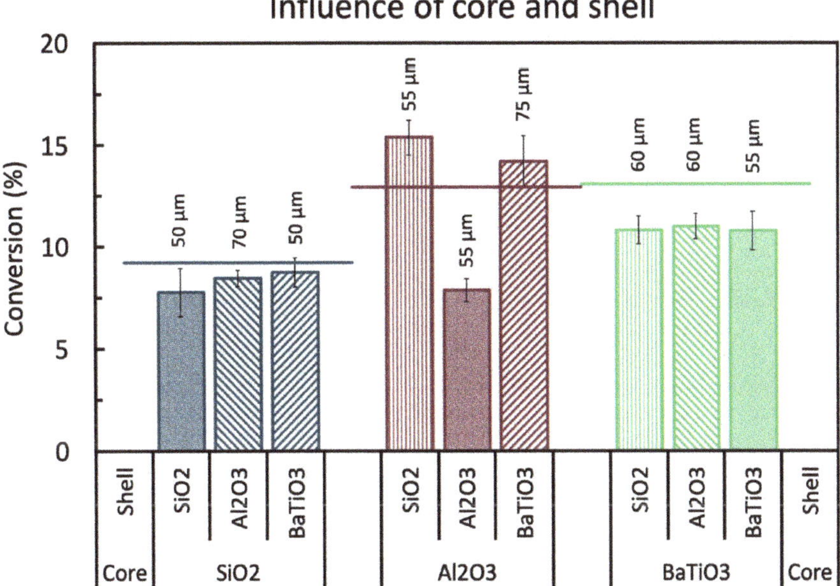

Figure 2. Influence of the core and shell material tested by approximately 50 µm thick shells applied on different core materials. All samples are compared to the pure spheres (solid line) with 1.6 to 1.8 mm size. All experiments are performed at the standard conditions. The exact values can be found in Table A1 in the Appendix A.

2.2.2. Influence of Core and Shell Material

By applying a thin shell of approximately 50 µm on a core, we can investigate the individual effect of the shell and core material on the performance of the core-shell sphere, while minimizing extra bulk effects that may be caused by the shell. The results of a combination of SiO_2, Al_2O_3, and $BaTiO_3$ in Figure 2 show that no clear order in shell performance can be observed, with the exact (lack of) impact depending on the core material. The results show that there is little to no difference in performance between the different shell materials when using SiO_2 cores or $BaTiO_3$ cores, while significant differences can be observed when shells are applied on Al_2O_3 cores. This suggests that there are no clear surface effects of the different shell materials in the case of SiO_2 cores or $BaTiO_3$ cores and/or that the shells are too thin to have a significant influence on the bulk effects of the whole sphere. However, the Al_2O_3 cores show a different story, because SiO_2 and $BaTiO_3$ shells slightly improve the base conversion; but the Al_2O_3-Al_2O_3 interaction, as already seen in the previous section, clearly has a detrimental effect.

Finally, Figure 2 illustrates the influence of the core material on the core-shell spheres. The general order of performance puts Al_2O_3 cores on top (with the exception of $Al_2O_3@Al_2O_3$), followed by $BaTiO_3$, and finally SiO_2. The exception of $Al_2O_3@Al_2O_3$ suggests that the Al_2O_3 powder (hypothesis (ii) from before) is the culprit of the bad performance and that it is not a core effect. From these results so far, we can conclude that optimal core-shell spheres can be designed by using a strong base material as a core, coupled to a complementary shell material. It is important to realize, that a strong core material is not necessarily also a strong shell material, and vice versa, due to the different (surface and bulk) properties that seem to play a role, as shown by the $Al_2O_3@Al_2O_3$ samples. This illustrates the high potential of core-shell spheres for optimal designed packing materials.

2.2.3. Influence of Shell Thickness

Finally, we take a look at the influence of the shell thickness, illustrated in Figure 1b–d with their exact values in Table A1 in Appendix A. It is clear that the exact effect of extra shell material is very different for all core and shell material combinations. SiO_2-based core-shell spheres, which have a low intrinsic performance, show to have opportunity for improvement as thicker Al_2O_3 and $BaTiO_3$ shells result in higher conversion. The added amount of shell material overcomes any effect of the SiO_2 cores, resulting in an almost linear increase in added performance.

This is, however, not the case with the Al_2O_3 and $BaTiO_3$ based core-shell spheres where no continuous increasing or decreasing behaviour is seen in function of shell thickness for either of the added shell chemistries. This suggests a strong interaction, or even competition, between the more prominently present but shielded surface and bulk effects of the core, and increasing surface and bulk effects of the shell. A combination effect of these properties seems to be present where for example adding more SiO_2 shifts the core-shell performance from pure core material (i.e., Al_2O_3 or $BaTiO_3$) more towards pure SiO_2, or the conversion of the $BaTiO_3@Al_2O_3$ spheres shifts between pure Al_2O_3 and $BaTiO_3$.

Curiously, SiO_2 shells do show an optimum thickness first, but it is unknown why exactly this 'poor behaving material' does this and at this particular thickness. Additionally, Al_2O_3 shells again show deviant behaviour, i.e., the performance does not rise as much on SiO_2 cores and there is low interaction with $BaTiO_3$ cores, suggesting that the coated powder has less activity than the shaped spheres (same powder, added binders)—see Section 2.2.2. These results show that optimizing core-shell spheres by thickness is possible, but the choice of core and shell material is very important and induces important additional properties. Moreover, it clearly shows that a good core material is not necessarily also a good candidate for a shell. The use of core-shell materials will most likely mainly influence the plasma discharge properties, as they are not necessarily catalytically active, but for specific reactions, they can also promote gas-surface reactions in case of a more porous available surface.

3. Discussion

A few reflections can be made based on the results from Sections 2.1 and 2.2. First of all, we did not find a core-shell sample with a significant performance improvement within this matrix of materials and shell thicknesses, as was also not intended. The results, however, did shed some light on the bulk and surface effects that different materials have on their performance in a DBD plasma reactor. The present data, however, do not allow us to determine the exact origins of their behaviour but do feature an impact of their relative contribution and thus importance. Better control over the material properties of the core-shell spheres is the next task to identify the underlying features of the results seen here. This requires a separate systematic, more elaborate study with much more controlled material synthesis based on the knowledge obtained in this work. Additionally, for the further development of tuned (catalytic) (core-shell) packing materials for plasma (catalysis) conversion processes, extra diagnostics would be needed to determine changes in the plasma electrical behaviour and plasma chemistry, such as optical emission spectroscopy or in-situ IR. This would allow to determine changes in the chemical species and to provide a more direct way for measuring the material effects on both the plasma gas composition and on the composition of the gas layer near, and adsorbed species on, the material surfaces (by in-situ spectroscopic measurements). This might elucidate the reactions that may occur at the surface, indicating any "catalysis". For instance, IR diagnostics in reflection mode on the surface, i.e., DRIFTS diffuse reflectance infrared Fourier transform spectroscopy) coupled to mass spectrometry (DRIFTS-MS) was developed by Stere et al. [35] for studying plasma-catalytic reactions through observation of changes in the species at the catalyst surface. This work was the first DRIFTS hybrid plasma (DBD) system reported in literature for studying the reaction mechanism during plasma catalysis. The same group also reported another interesting in-situ diagnostic [36] for studying the role of plasma in heating on the catalyst structure, using X-ray absorption fine structure (XAFS). Azzolina-Jury and Thibault-Starzyk also applied IR diagnostics, to obtain time-resolved in-situ spectroscopic data, directly providing information about adsorbed species under plasma exposure [37,38]. Note, however, that these techniques are most sensitive to species with large surface density, which are not necessarily the most reactive species. Furthermore, some reactive intermediate species may not be detected, depending on the time resolution. In addition, fast imaging by intensified Charge Coupled Device (iCCD) cameras can be applied to study the plasma behaviour in contact with catalyst materials, and in particular the characteristics of plasma streamers propagating over catalyst surfaces [39–41]. Finally, plasma dynamic experiments of the spheres in a simplified and standardized packed bed set-up, such as proposed by Butterworth and Allen [42], can provide information on the changes in plasma discharging behaviour for different core-shell materials.

Besides being able to tune and optimize the bulk and surface effects of packing materials through core-shell spheres, the macroporous shell structure forms interesting opportunities for catalyst doping. It was predicted by modelling that plasma streamers can only penetrate into pores with a size larger than the Debye length [33], which is typically a few 100 nm, depending on the gas and other operating conditions. Hence, full dispersion of a catalytic compound in the entire (microporous) sphere can be a waste of valuable catalyst material. The thin shell layers produced in our work can form the perfect substrate for catalyst doping for more optimal plasma-catalyst interaction. Bulk effects of the core-shell sphere can optimize the plasma and reactive species generation via its physical impact, while the doped shell material can create the optimal accessible surface needed for catalyst reactions, with the possibility to enhance reaction pathways via the plasma-excited species.

Additionally, we recently investigated the reaction rate coefficients and equilibrium constants of CO_2 dissociation, CH_4 reforming, and dry reforming of methane under the influence of power, pressure, discharge gap size, and packing materials [29]. By testing packed-bed reactors over an extended residence time range, we discovered that packing materials can individually change the overall reaction rate coefficient, as well as the equilibrium position. By checking only one particular condition, a lot of possible information is lost about the exact effect of a certain type of packing material on the kinetics. Therefore, further endeavours in (catalytic) packing material development will benefit from using this type of analysis to obtain more detailed knowledge.

Finally, Figure 1 also displays the energy efficiency of the benchmark results and all core-shell samples. It shows the same trends as the conversion, since the energy efficiency is linearly proportional to the conversion when the power and flow rate are constant. The reference energy efficiency of the empty reactor was found to be only 3% for both cases, i.e., the same flow rate and same residence time as the packed reactors. Adding the pure un-coated spheres can enhance the energy efficiency, depending on the material and size combination, up to 4.1% for 2.0 to 2.24 mm Al_2O_3 spheres. Within the core-shell samples, only the 55 μm $SiO_2@Al_2O_3$ and 250 μm $SiO_2@BaTiO_3$ can slightly enhance this energy efficiency further, although all within the same error bars. This matches the maximum values we have found before for CO_2 dissociation in our previous paper on packed-bed (micro) DBD reactors [5].

4. Materials and Methods

4.1. Core-Shell Spheres

The core-shell spheres were synthesized by means of spray coating. With this method, a suspension of the desired shell material is sprayed on the cores in fine droplets, which stick to the surface, and as the solvent evaporates, it leaves a fraction of powder behind. Spraying while turning the core materials in a pan, slowly builds up a full layer over time, that can grow to a desired thickness. A calcination process was applied to fix the layer in place and remove the organic components.

The formulation of the coating suspension is the same for all core-shell combinations and was based on the coating slurry of Lefevere et al. for their 3DFD structures [43]. A mixture of distilled water, methyl cellulose as a temporary organic binder (15 cP, Sigma-Aldrich, Overijse, Belgium), and colloidal silica as a permanent binder (LUDOX HS-40, Sigma-Aldrich) was used to disperse and suspend the powderous shell material. The final composition was 1 wt% methyl cellulose, 1 wt% LUDOX, and 30 wt% shell powder. Powders made of SiO_2 (0–10 μm, Sigmund Lindner, Warmensteinach, Germany), α-Al_2O_3 (400 nm A 16 SG, Almatis, Rotterdam, The Netherlands), and $BaTiO_3$ (200 nm, Inframat, Manchester, Connecticut, United States) were used as the shell materials as received. For each new batch, 100 mL of bare spheres, used as the cores, were placed in a rotating bowl with agitation fins to tumble the spheres around and ensure an as even as possible coating of all spheres and sphere surface. Dense SiO_2 (Type-S, Sigmund Lindner), α-Al_2O_3 (custom made at VITO, Mol, Belgium) [34], and $BaTiO_3$ (Catal Ltd., Sheffield United Kingdom) spheres with a size of 1.6 to 1.8 mm were used as cores. The suspension was coated on the rotating spheres by a compressed air driven spray gun, and dried at the same time with a hot air gun. The green core-shell spheres were calcined for 4 h at 650 °C with a heating ramp of 2 °C/min. The four shell thicknesses where aimed at being 50, 100, 150, and 200 μm. Maximum layers of approximately 100 μm were applied at a time. The 150 and 200 μm coatings were done in two steps with an intermediate calcination step. In practice, deviant shell thicknesses will be obtained due to the unpredictable suspension losses during the coating process, i.e., premature drying of the sprayed droplets and abrasion losses during tumbling. The shell thickness was analysed after calcination by embedding the spheres in an epoxy resin (ClaroCit, Struers, Maassluis, The Netherlands), sanding them down until halfway the spheres, being imaged by 10x microscope (XploRa plus, Horiba Scientific, Villeneuve-d'Ascq, France), and being analysed by ImageJ (version 1.52, National Institute for Health, United States). A schematic representation of the spray coating set-up and an example of four layer thicknesses of $BaTiO_3@SiO_2$ core-shell spheres are shown in Appendices B and C (Figures A1 and A2, respectively).

4.2. Experimental Set-Up

A cylindrical DBD reactor was used, as schematically shown in Figure 3. The reactor consists of an alumina dielectric tube, with a 22 mm outer diameter and inner diameter of 17 mm, which serves as the dielectric layer and outer wall of the reaction zone. A stainless-steel rod, with an 8 mm outer diameter, was placed in the centre of the tube to serve as the inner, grounded electrode and inner wall of the reactor, resulting in a discharge gap of 4.5 mm. A stainless-steel mesh, with a length of 100 mm, was wrapped around the outside of the alumina tube, acting as the high voltage electrode, resulting in a total reaction volume of 17.67 mL. The reaction volume was filled with either the reference dense spheres or synthesized core-shell spheres, which were held in place with a layer of glass wool (superfine 8 to 50 μm, Glaswarenfabrik Karl Hecht, Sondheim vor der Rhön, Germany) on both sides. The reference spheres are the same cores as in Section 4.1, in two size ranges, being 1.6–1.8 mm and 2.0–2.24 mm.

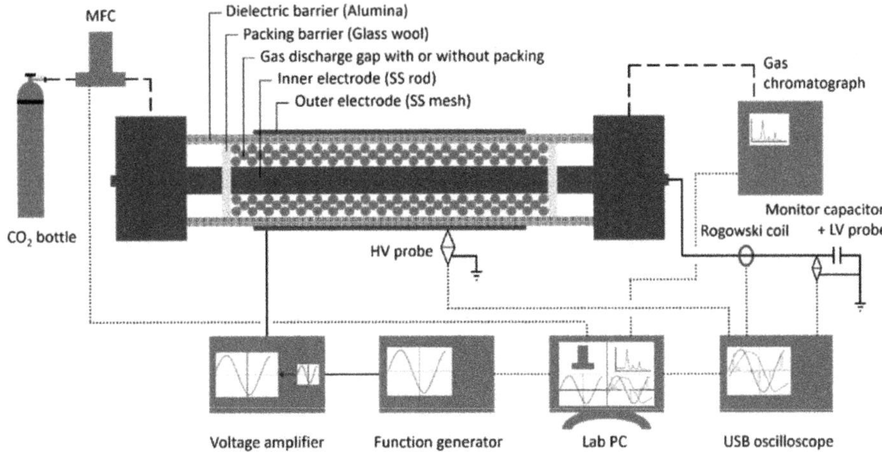

Figure 3. Packed bed DBD plasma reactor used in this work, along with analytical equipment (reprinted from [5] with permission from Elsevier).

A pure CO_2 stream was fed to the reactor via a mass flow controller. A flow rate of 38.98 mL/min was used for both empty and packed reactors, and in addition, a flow rate of 75.35 mL/min was used for the empty reactor to test the performance at equal residence time as the packed reactor at 38.98 mL/min (i.e., 14.07 s for a modelled packing efficiency of 48.27% [29]).

The reaction products were analysed by a gas chromatograph (Compact GC, Interscience, Breda, The Netherlands) with pressure-less sampling. This GC features a thermal conductivity detector (TCD) channel, able to measure the CO and O_2 composition as one peak and the CO_2 composition separated by an Rt-Q-Bond column. No significant amounts of ozone or carbon deposition were detected. The CO_2 conversion was defined as:

$$X_{GC,CO_2} = \frac{\dot{CO}_{2in} - \dot{CO}_{2out}}{\dot{CO}_{2in}},\tag{1}$$

where $\dot{CO}_2$ is the molar flow rate of CO_2. A correction factor was applied to compensate for gas expansion in a pressure-less set-up to obtain the actual conversion X_{CO_2} [44,45]:

$$X_{GC,CO_2} = 1 - \left(\frac{1 - X_{CO_2}}{1 + \frac{X_{CO_2}}{2}}\right) \Leftrightarrow X_{CO_2} = \frac{2\,X_{GC,CO_2}}{3 - X_{GC,CO_2}}.\tag{2}$$

The power to the reactor was provided by a high voltage amplifier (model 20/20C-HS, TREK, Lockport, NY, USA), which was driven by a PC controlled function generator (AFG 2021, Tektronix, Berkshire, UK) at a frequency of 3 kHz. The applied voltage, current, and displaced charge were monitored by a high voltage probe (P6015A, Tektronix), a current transformer (model 4100, Pearson Electronics, Palo Alto, CA, USA), and a low voltage probe (TA150, Pico Technology, St. Neots, UK) paired with a monitor capacitor (10 nF), respectively. All signals were recorded with a digital oscilloscope (Picoscope 6402D, Pico Technology). The power was calculated during a number (n) of consecutive periods (T) according to:

$$P = \frac{1}{nT}\int_0^{nT} U(t)I(t)dt.\tag{3}$$

The power was maintained at 30 W by adjusting the voltage of the function generator. Finally, the energy efficiency was calculated based on the theoretical and consumed energy as:

$$\eta = \frac{\Delta H_r \, X_{CO_2} \, \dot{V}}{P \, V_m},\tag{4}$$

where ΔH_r is the reaction enthalpy of CO_2 dissociation (279.8 kJ/mol), $\dot{V}$ is the volumetric flow rate, P is the plasma power, and V_m is the molar gas volume (22.4 L/mol). The ratio of the plasma power over the volumetric flow rate is also known as the specific energy input or *SEI*:

$$SEI = \frac{P}{\dot{V}},\tag{5}$$

4.3. Experimental Method

Each experiment was started with a freshly packed, cooled-down reactor and operated for 40 min to achieve steady state conversion. The amplitude of the input voltage was continuously adjusted to deliver the desired power of 30 W during this stabilisation, which was followed by the GC and Lissajous measurements.

Each packing material was tested in threefold with four GC and Lissajous measurements for statistical analysis. The error on the resulting average is defined as:

$$error = \pm S_n \frac{T_s(p, n_s)}{\sqrt{n_s}},\tag{6}$$

where S_n is the sample standard deviation of the measurements, n_s is the sample size being 12, and T_s is the two-tailed inverse of the Student t-distribution for sample size n_s and probability p set at 95%.

The standard operating conditions used in this work were a pure CO_2 stream with a flow rate of 38.98 mL/min, performed at 30 W (3 kHz), and 1 bar in a discharge gap of 4.5 mm. This results in an average residence time of 14.07 s and a specific energy input (SEI) of 46.18 J/mL in the packed reactors. Some of these parameters were varied, as specified in the results section.

5. Conclusions

In this paper we investigated the potential of core-shell structured spheres as packing materials in a DBD reactor, for plasma-based CO_2 conversion. Core-shell spheres have the potential to be tailored to a specific reaction, requiring weak/strong bulk/surface effects, potentially in combination with a catalytically active material for optimal performance. First of all, we found that applying a thin shell layer of approximately 50 µm of the same material as the core significantly reduces the performance of the packing material, indicating that the shell might mask the positive effect of the core and/or induced negative effects due to certain shell properties. Combining different materials showed various interactions between the core and shell material, affecting the conversion. Al_2O_3 was found to be the best core material, followed by $BaTiO_3$ and SiO_2, in agreement with the behaviour of the pure spheres. It was also found that all three shell materials perform equal in low amounts (thin shell), with the exception of $Al_2O_3@Al_2O_3$, and that they are not able to provide any significant improvement. A strong mixing behaviour is seen where more active shell materials can improve weak core materials, but will have to compete against strong core materials to show their effect on the performance.

Our results show that surface and bulk effects can have different influences on the performance of the spheres in a plasma reactor. A strong core material is not necessarily also a strong shell material, and vice versa, due to the different (surface and bulk) properties that seem to play a role; as shown by the $Al_2O_3@Al_2O_3$ samples. This illustrates a great potential, as using core-shell spheres can provide us with the possibility of tuning the packing properties more closely to the application. Furthermore, the thin porous nature of the shell offers possibilities to dope a packing material with just the right amount of catalyst for plasma catalysis, compared to fully porous supports, where catalyst material could be wasted, as the plasma cannot penetrate deeply into pores (with a minimum diameter of a few 100 nm).

Author Contributions: Conceptualization, Y.U.; data curation, Y.U.; investigation, Y.U.; methodology, Y.U.; supervision, V.M., P.C. and A.B.; writing—original draft, Y.U.; writing—review and editing, V.M., P.C. and A.B. All authors have read and agreed to the published version of the manuscript.

Funding: We acknowledge financial support from the European Fund for Regional Development through the cross-border collaborative Interreg V program Flanders-the Netherlands (project EnOp), the Fund for Scientific Research (FWO; grant number: G.0254.14N) and an IOF-SBO (SynCO2Chem) project from the University of Antwerp.

Acknowledgments: We want to thank Jasper Lefevre (VITO) for assistance in the development of the coating suspension for the core-shell spheres.

Conflicts of Interest: The authors declare no conflict of interest.

Appendix A. Raw Data of Figure 1

Table A1. Conversion and energy efficiency of (i) the empty DBD reactor, both at the same flow rate and residence time (RT) as the packed bed reactor, and (ii) of all samples used in this work (including pure uncoated spheres and the various combinations of core-shell spheres) as shown in Figures 1 and 2.

Sample	Size (mm)/Shell Thickness (μm)	Conversion (%)	Energy Efficiency (%)
Empty (=Flow)	/	11 ± 1	3.0 ± 0.3
Empty (=RT)	/	6.4 ± 0.8	3.2 ± 0.5
SiO_2	1.6–1.8	9.8 ± 0.3	2.7 ± 0.1
Al_2O_3	1.6–1.8	13 ± 1	3.5 ± 0.3
$BaTiO_3$	1.6–1.8	13 ± 1	3.5 ± 0.3
SiO_2	2.0–2.24	9.2 ± 0.8	2.5 ± 0.2
Al_2O_3	2.0–2.24	15.2 ± 0.9	4.1 ± 0.2
$BaTiO_3$	2.0–2.24	10.9 ± 0.7	3.0 ± 0.2
$SiO_2@SiO_2$	50	8 ± 1	2.1 ± 0.3
$Al_2O_3@SiO_2$	70	8.5 ± 0.4	2.2 ± 0.1
$Al_2O_3@SiO_2$	140	9.7 ± 0.9	2.7 ± 0.2
$Al_2O_3@SiO_2$	185	10.2 ± 0.6	2.7 ± 0.2
$Al_2O_3@SiO_2$	250	9.9 ± 0.9	2.7 ± 0.2
$BaTiO_3@SiO_2$	50	8.7 ± 0.7	2.4 ± 0.2
$BaTiO_3@SiO_2$	125	9.8 ± 0.8	2.7 ± 0.2
$BaTiO_3@SiO_2$	180	11.2 ± 0.8	3.0 ± 0.2
$BaTiO_3@SiO_2$	225	11.9 ± 0.9	3.2 ± 0.2
$SiO_2@Al_2O_3$	55	15.4 ± 0.9	4.2 ± 0.2
$SiO_2@Al_2O_3$	100	10.8 ± 0.8	2.9 ± 0.2
$SiO_2@Al_2O_3$	290	11.8 ± 0.6	3.2 ± 0.2
$SiO_2@Al_2O_3$	405	12.0 ± 0.7	3.3 ± 0.2
$Al_2O_3@Al_2O_3$	55	7.9 ± 0.6	2.2 ± 0.2
$BaTiO_3@Al_2O_3$	75	14 ± 1	3.8 ± 0.3
$BaTiO_3@Al_2O_3$	90	11.3 ± 0.7	3.0 ± 0.2
$BaTiO_3@Al_2O_3$	160	11 ± 1	3.2 ± 0.3
$BaTiO_3@Al_2O_3$	230	14.3 ± 0.4	3.8 ± 0.1

Table A1. *Cont.*

Sample	Size (mm)/Shell Thickness (µm)	Conversion (%)	Energy Efficiency (%)
SiO_2@$BaTiO_3$	60	10.8 ± 0.7	2.9 ± 0.2
SiO_2@$BaTiO_3$	250	16 ± 1	4.3 ± 0.3
SiO_2@$BaTiO_3$	370	13.0 ± 0.8	3.5 ± 0.2
SiO_2@$BaTiO_3$	465	12 ± 1	3.2 ± 0.3
Al_2O_3@$BaTiO_3$	60	11.0 ± 0.6	3.0 ± 0.2
Al_2O_3@$BaTiO_3$	125	8.3 ± 0.8	2.3 ± 0.2
Al_2O_3@$BaTiO_3$	165	10 ± 1	2.8 ± 0.3
Al_2O_3@$BaTiO_3$	235	10.2 ± 0.4	2.8 ± 0.1
$BaTiO_3$@$BaTiO_3$	55	10.8 ± 0.9	2.9 ± 0.2

Appendix B. Schematic Representation of the Spray Coating Set-Up Used in This Work

Figure A1 shows a schematic representation of the spray coating set-up used in this work. It is an in-house built pan coating set-up comprised of a rotating pan with agitation fins added to the inside to disturb the rolling spheres into tumbling over each other. The coating suspension is added by a gravity fed spray gun operated with compressed dry air at 1–1.5 barg. The suspension is gradually added and sprayed on the spheres in the pan, while the remaining fraction is left in a beaker on a stirring plate. The pan and contents are continuously heated by a hot air gun operated at maximum heat but medium air flow rate, to maximize the heating capacity but to minimize deflection of the sprayed droplets away from the tumbling spheres.

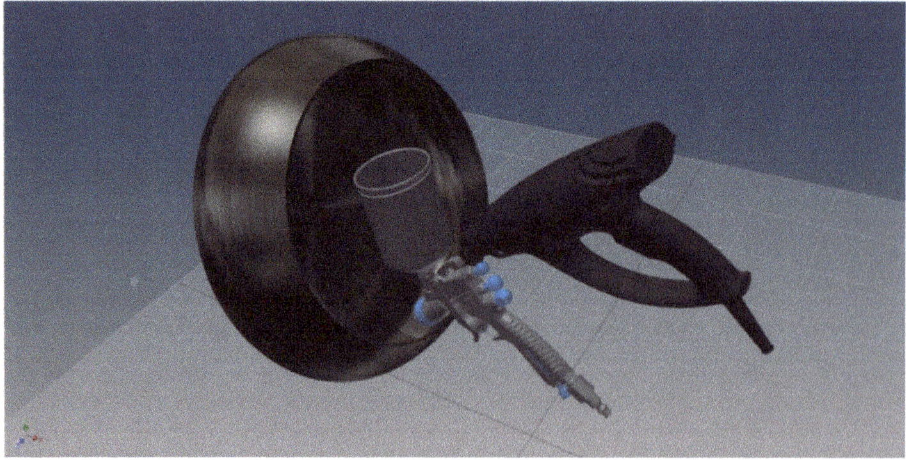

Figure A1. Schematic representation of the spray coating set-up used in this work.

Appendix C. Example of Four Layer Thicknesses of BaTiO₃@SiO₂ Core-Shell Spheres

Figure A2 shows an example of the four layer thicknesses of $BaTiO_3$ applied on the 1.6 to 1.8 mm SiO_2 cores. These images are obtained by embedding samples of about 25 spheres of the different core-shell spheres in an epoxy resin and sanding it down to about half-way the spheres. The spheres were then imaged by light microscopy by overlapping multiple exposures (hence the visible rectangular pattern). The average layer thicknesses were measured by ImageJ analysis.

Uniform coverage of the entire spheres was obtained for every layer thickness but some shell roughness is present due to the tumbling spray coating method.

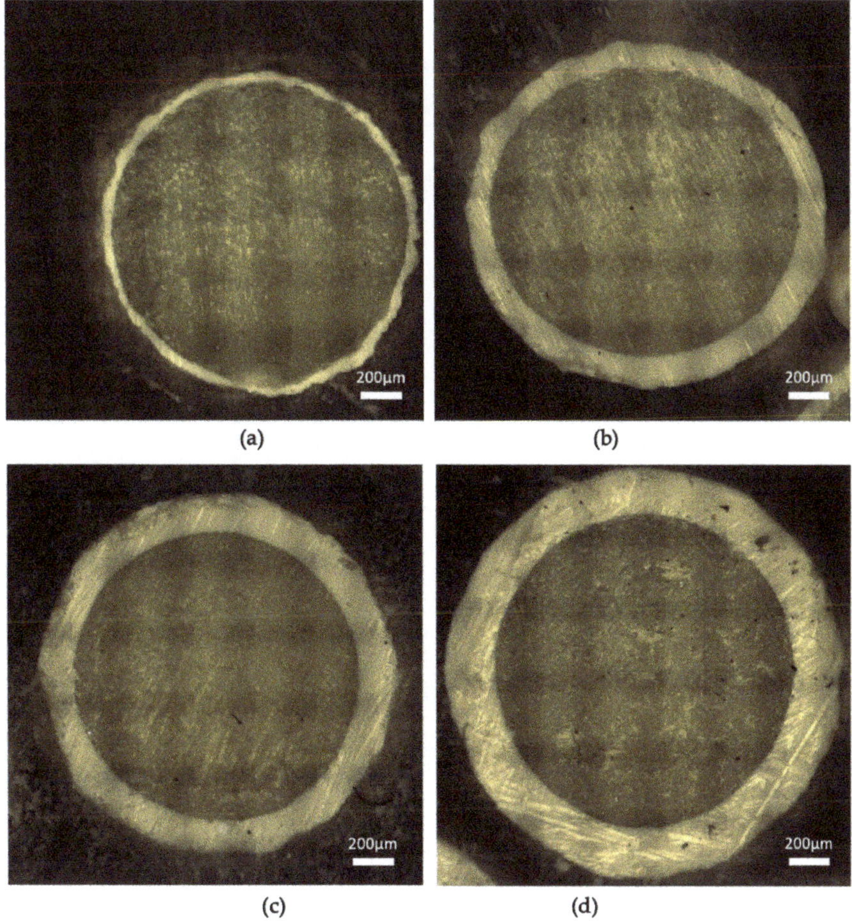

Figure A2. Example of four layer thicknesses of $BaTiO_3$@SiO_2 core-shell spheres; (**a**): 50 μm; (**b**): 130 μm; (**c**): 180 μm; (**d**): 230 μm.

References

1. Kim, H.-H. Nonthermal plasma processing for air-pollution control: A historical review, current issues, and future prospects. *Plasma Process. Polym.* **2004**, *1*, 91–110. [CrossRef]
2. Van Durme, J.; Dewulf, J.; Leys, C.; Van Langenhove, H. Combining non-thermal plasma with heterogeneous catalysis in waste gas treatment: A review. *Appl. Catal. B Environ.* **2008**, *78*, 324–333. [CrossRef]
3. Kogelschatz, U. Dielectric-barrier discharges: Their history, discharge physics, and industrial applications. *Plasma Chem. Plasma Process.* **2003**, *23*, 1–46. [CrossRef]
4. Snoeckx, R.; Bogaerts, A. Plasma technology—A novel solution for CO_2 conversion? *Chem. Soc. Rev.* **2017**, *46*, 5805–5863. [CrossRef] [PubMed]
5. Uytdenhouwen, Y.; Van Alphen, S.; Michielsen, I.; Meynen, V.; Cool, P.; Bogaerts, A. A packed-bed DBD micro plasma reactor for CO_2 dissociation: Does size matter? *Chem. Eng. J.* **2018**, *348*, 557–568. [CrossRef]
6. Michielsen, I.; Uytdenhouwen, Y.; Pype, J.; Michielsen, B.; Mertens, J.; Reniers, F.; Meynen, V.; Bogaerts, A. CO_2 dissociation in a packed bed DBD reactor: First steps towards a better understanding of plasma catalysis. *Chem. Eng. J.* **2017**, *326*, 477–488. [CrossRef]
7. Butterworth, T.; Elder, R.; Allen, R. Effects of particle size on CO_2 reduction and discharge characteristics in a packed bed plasma reactor. *Chem. Eng. J.* **2016**, *293*, 55–67. [CrossRef]

8. Neyts, E.C.; Bogaerts, A. Understanding plasma catalysis through modelling and simulation—A review. *J. Phys. D Appl. Phys.* **2014**, *47*, 224010. [CrossRef]

9. Van Laer, K.; Bogaerts, A. How bead size and dielectric constant affect the plasma behaviour in a packed bed plasma reactor: A modelling study. *Plasma Sources Sci. Technol.* **2017**, *26*, 085007. [CrossRef]

10. Whitehead, J.C. Plasma—Catalysis: The known knowns, the known unknowns and the unknown unknowns. *J. Phys. D Appl. Phys.* **2016**, *49*, 243001. [CrossRef]

11. Michielsen, I.; Uytdenhouwen, Y.; Bogaerts, A.; Meynen, V. Altering Conversion and Product Selectivity of Dry Reforming of Methane in a Dielectric Barrier Discharge by Changing the Dielectric Packing Material. *Catalysts* **2019**, *9*, 51. [CrossRef]

12. Roosta, Z.; Izadbakhsh, A.; Sanati, A.M.; Osfouri, S. Synthesis and evaluation of NiO@MCM-41 core–shell nanocomposite in the CO_2 reforming of methane. *J. Porous Mater.* **2018**, *25*, 1135–1145. [CrossRef]

13. Bao, J.; He, J.; Zhang, Y.; Yoneyama, Y.; Tsubaki, N. A core/shell catalyst produces a spatially confined effect and shape selectivity in a consecutive reaction. *Angew. Chemie Int. Ed.* **2008**, *47*, 353–356. [CrossRef] [PubMed]

14. Zhong, C.J.; Maye, M.M. Core-shell assembled nanoparticles as catalysts. *Adv. Mater.* **2001**, *13*, 1507–1511. [CrossRef]

15. Joo, S.H.; Park, J.Y.; Tsung, C.K.; Yamada, Y.; Yang, P.; Somorjai, G.A. Thermally stable Pt/mesoporous silica core-shell nanocatalysts for high-temperature reactions. *Nat. Mater.* **2009**, *8*, 126–131. [CrossRef] [PubMed]

16. Yu, L.; Ni, C.; Grist, S.M.; Bayly, C.; Cheung, K.C. Alginate core-shell beads for simplified three-dimensional tumor spheroid culture and drug screening. *Biomed. Microdevices* **2015**, *17*, 33. [CrossRef]

17. Ramli, R.A.; Laftah, W.A.; Hashim, S. Core-shell polymers: A review. *RSC Adv.* **2013**, *3*, 15543–15565. [CrossRef]

18. He, Z.; Tu, R.; Katsui, H.; Goto, T. Synthesis of SiC/SiO_2 core-shell powder by rotary chemical vapor deposition and its consolidation by spark plasma sintering. *Ceram. Int.* **2013**, *39*, 2605–2610. [CrossRef]

19. Yang, X.H.; Fu, H.T.; Wong, K.; Jiang, X.C.; Yu, A.B. Hybrid Ag@TiO_2 core-shell nanostructures with highly enhanced photocatalytic performance. *Nanotechnology* **2013**, *24*, 415601. [CrossRef]

20. Luc, W.; Collins, C.; Wang, S.; Xin, H.; He, K.; Kang, Y.; Jiao, F. Ag-sn bimetallic catalyst with a core-shell structure for CO_2 reduction. *J. Am. Chem. Soc.* **2017**, *139*, 1885–1893. [CrossRef]

21. Palugan, L.; Cerea, M.; Zema, L.; Gazzaniga, A.; Maroni, A. Coated pellets for oral colon delivery. *J. Drug Deliv. Sci. Technol.* **2015**, *25*, 1–15. [CrossRef]

22. Varshosaz, J.; Emami, J.; Tavakoli, N.; Minaiyan, M.; Rahmani, N.; Dorkoosh, F. Development and Evaluation of a Novel Pellet-Based Tablet System for Potential Colon Delivery of Budesonide. *J. Drug Deliv.* **2012**, *2012*, 1–7. [CrossRef] [PubMed]

23. Liu, J.Y.; Zhang, X.X.; Huang, H.Y.; Lee, B.J.; Cui, J.H.; Cao, Q.R. Esomeprazole magnesium enteric-coated pellet-based tablets with high acid tolerance and good compressibility. *J. Pharm. Investig.* **2018**, *48*, 341–350. [CrossRef]

24. Hampel, N.; Bück, A.; Peglow, M.; Tsotsas, E. Continuous pellet coating in a Wurster fluidized bed process. *Chem. Eng. Sci.* **2013**, *86*, 87–98. [CrossRef]

25. Zheng, X.; Tan, S.; Dong, L.; Li, S.; Chen, H. LaNiO$_3$@SiO_2 core-shell nano-particles for the dry reforming of CH_4 in the dielectric barrier discharge plasma. *Int. J. Hydrog. Energy* **2014**, *39*, 11360–11367. [CrossRef]

26. Zheng, X.; Tan, S.; Dong, L.; Li, S.; Chen, H. Plasma-assisted catalytic dry reforming of methane: Highly catalytic performance of nickel ferrite nanoparticles embedded in silica. *J. Power Sources* **2015**, *274*, 286–294. [CrossRef]

27. Hong, J.; Aramesh, M.; Shimoni, O.; Seo, D.H.; Yick, S.; Greig, A.; Charles, C.; Prawer, S.; Murphy, A.B. Plasma Catalytic Synthesis of Ammonia Using Functionalized-Carbon Coatings in an Atmospheric-Pressure Non-equilibrium Discharge. *Plasma Chem. Plasma Process.* **2016**, *36*, 917–940. [CrossRef]

28. Van Laer, K.; Bogaerts, A. Influence of gap size and dielectric constant of the packing material on the plasma behaviour in a packed bed DBD reactor: A fluid modelling study. *Plasma Process. Polym.* **2017**, *14*, 1600129. [CrossRef]

29. Uytdenhouwen, Y.; Bal, K.M.; Michielsen, I.; Neyts, E.C.; Meynen, V.; Cool, P.; Bogaerts, A. How process parameters and packing materials tune chemical equilibrium and kinetics in plasma-based CO_2 conversion. *Chem. Eng. J.* **2019**, *372*, 1253–1264. [CrossRef]

30. Lefevere, J.; Protasova, L.; Mullens, S.; Meynen, V. 3D-printing of hierarchical porous ZSM-5: The importance of the binder system. *Mater. Des.* **2017**, *134*, 331–341. [CrossRef]
31. Lefevere, J.; Mullens, S.; Meynen, V. The impact of formulation and 3D-printing on the catalytic properties of ZSM-5 zeolite. *Chem. Eng. J.* **2018**, *349*, 260–268. [CrossRef]
32. Vajglová, Z.; Kumar, N.; Mäki-Arvela, P.; Eränen, K.; Peurla, M.; Hupa, L.; Murzin, D.Y. Effect of Binders on the Physicochemical and Catalytic Properties of Extrudate-Shaped Beta Zeolite Catalysts for Cyclization of Citronellal. *Org. Process Res. Dev.* **2019**, *23*, 2456–2463. [CrossRef]
33. Zhang, Q.-Z.; Bogaerts, A. Propagation of a plasma streamer in catalyst pores. *Plasma Sources Sci. Technol.* **2018**, *27*, 035009. [CrossRef]
34. Pype, J.; Michielsen, B.; Seftel, E.M.; Mullens, S.; Meynen, V. Development of alumina microspheres with controlled size and shape by vibrational droplet coagulation. *J. Eur. Ceram. Soc.* **2017**, *37*, 189–198. [CrossRef]
35. Stere, C.E.; Adress, W.; Burch, R.; Chansai, S.; Goguet, A.; Graham, W.G.; Hardacre, C. Probing a non-thermal plasma activated heterogeneously catalyzed reaction using in situ DRIFTS-MS. *ACS Catal.* **2015**, *5*, 956–964. [CrossRef]
36. Gibson, E.K.; Stere, C.E.; Curran-McAteer, B.; Jones, W.; Cibin, G.; Gianolio, D.; Goguet, A.; Wells, P.P.; Catlow, C.R.A.; Collier, P.; et al. Probing the Role of a Non-Thermal Plasma (NTP) in the Hybrid NTP Catalytic Oxidation of Methane. *Angew. Chemie Int. Ed.* **2017**, *56*, 9351–9355. [CrossRef]
37. Azzolina-Jury, F. Novel boehmite transformation into γ-alumina and preparation of efficient nickel base alumina porous extrudates for plasma-assisted CO2 methanation. *J. Ind. Eng. Chem.* **2019**, *71*, 410–424. [CrossRef]
38. Azzolina-Jury, F.; Thibault-Starzyk, F. Mechanism of Low Pressure Plasma-Assisted CO_2 Hydrogenation Over Ni-USY by Microsecond Time-resolved FTIR Spectroscopy. *Top. Catal.* **2017**, *60*, 1709–1721. [CrossRef]
39. Kim, H.-H.; Teramoto, Y.; Ogata, A. Time-resolved imaging of positive pulsed corona-induced surface streamers on TiO_2 and γ-Al_2O_3 -supported Ag catalysts. *J. Phys. D Appl. Phys.* **2016**, *49*, 459501. [CrossRef]
40. Kim, H.H.; Teramoto, Y.; Ogata, A.; Kang, W.S.; Hur, M.; Song, Y.H. Negative surface streamers propagating on TiO_2 and γ-Al_2O_3-supported Ag catalysts: ICCD imaging and modeling study. *J. Phys. D Appl. Phys.* **2018**, *51*, 244006. [CrossRef]
41. Wang, W.; Kim, H.H.; Van Laer, K.; Bogaerts, A. Streamer propagation in a packed bed plasma reactor for plasma catalysis applications. *Chem. Eng. J.* **2018**, *334*, 2467–2479. [CrossRef]
42. Butterworth, T.; Allen, R.W.K. Plasma-catalyst interaction studied in a single pellet DBD reactor: Dielectric constant effect on plasma dynamics. *Plasma Sources Sci. Technol.* **2017**, *26*, 065008. [CrossRef]
43. Lefevere, J.; Gysen, M.; Mullens, S.; Meynen, V.; Van Noyen, J. The benefit of design of support architectures for zeolite coated structured catalysts for methanol-to-olefin conversion. *Catal. Today* **2013**, *216*, 18–23. [CrossRef]
44. Pinhão, N.; Moura, A.; Branco, J.B.; Neves, J. Influence of gas expansion on process parameters in non-thermal plasma plug-flow reactors: A study applied to dry reforming of methane. *Int. J. Hydrog. Energy* **2016**, *41*, 9245–9255. [CrossRef]
45. Snoeckx, R.; Heijkers, S.; Van Wesenbeeck, K.; Lenaerts, S.; Bogaerts, A. CO_2 conversion in a dielectric barrier discharge plasma: N_2 in the mix as a helping hand or problematic impurity? *Energy Environ. Sci. Energy Environ. Sci.* **2016**, *9*, 999–1011. [CrossRef]

 *catalysts*

Article

The Effect of Packing Material Properties on Tars Removal by Plasma Catalysis

Richard Cimerman [1],*, Mária Cíbiková [1], Leonid Satrapinskyy [2] and Karol Hensel [1]

[1] Department of Astronomy, Physics of the Earth and Meteorology, Faculty of Mathematics, Physics and Informatics, Comenius University, Mlynská Dolina F2, 842 48 Bratislava, Slovakia; cibikova17@uniba.sk (M.C.); karol.hensel@fmph.uniba.sk (K.H.)

[2] Department of Experimental Physics, Faculty of Mathematics, Physics and Informatics, Comenius University, Mlynská Dolina F2, 842 48 Bratislava, Slovakia; leonid.satrapinskyy@fmph.uniba.sk

* Correspondence: richard.cimerman@fmph.uniba.sk

Received: 15 November 2020; Accepted: 14 December 2020; Published: 17 December 2020

Abstract: Plasma catalysis has been utilized in many environmental applications for removal of various hydrocarbons including tars. The aim of this work was to study the tars removal process by atmospheric pressure DBD non-thermal plasma generated in combination with packing materials of various composition and catalytic activity (TiO_2, $Pt/\gamma Al_2O_3$, $BaTiO_3$, γAl_2O_3, ZrO_2, glass beads), dielectric constant (5–4000), shape (spherical and cylindrical pellets and beads), size (3–5 mm in diameter, 3–8 mm in length), and specific surface area (37–150 m^2/g). Naphthalene was chosen as a model tar compound. The experiments were performed at a temperature of 100 °C and a naphthalene initial concentration of approx. 3000 ppm, i.e., under conditions that are usually less favorable to achieve high removal efficiencies. For a given specific input energy of 320 J/L, naphthalene removal efficiency followed a sequence: TiO_2 > $Pt/\gamma Al_2O_3$ > ZrO_2 > γAl_2O_3 > glass beads > $BaTiO_3$ > plasma only. The efficiency increased with the increasing specific surface area of a given packing material, while its shape and size were also found to be important. By-products of naphthalene decomposition were analyzed by means of FTIR spectrometry and surface of packing materials by SEM analysis.

Keywords: non-thermal plasma; dielectric barrier discharge; plasma catalysis; naphthalene removal; FTIR spectrometry; SEM analysis; TiO_2; $Pt/\gamma Al_2O_3$; ZrO_2; $BaTiO_3$

1. Introduction

In recent years, non-thermal plasma (NTP) generated by atmospheric pressure electric discharges in a combination with catalysis, i.e., plasma catalysis, is gaining an increasing interest as it offers a high plasma reactivity along with a high catalytic selectivity [1,2]. In addition, plasma catalysis is often characterized by synergistic effects, when the combined effect of plasma with catalysis is usually stronger than their individual effects [3–5]. Although the exact underlying mechanisms of plasma catalysis synergy are still unclear, several hypotheses based on experimental observations have been proposed: plasma-induced adsorption/desorption, activation of lattice oxygen, lowering the activation barriers, generation of electron–hole pairs by the absorption of UV radiation, direct interaction of gaseous radicals with the catalyst surface and the adsorbed molecules, penetration of plasma reactive species into the pores of the catalyst, local heating (hot spots), etc. [2,4,6,7]. The synergy may lead to higher conversion of reactants, higher selectivity and yield of desired products as well as higher energy efficiency of the process [8,9]. Thanks to this, plasma catalysis has found utilization in many environmental applications particularly in air pollution control technologies, including removal of nitrogen oxides (NOx) [10–13], volatile organic compounds (VOCs) [14–17], polycyclic aromatic hydrocarbons (PAHs) [18], and conversion of carbon dioxide CO_2 [8,19,20].

The utilization of NTP in the PAH removal process has been investigated by many authors and extensive and comprehensive reviews about the topic have been published in several papers [18,21–23]. Although the PAH removal by plasma catalysis was investigated much less frequently in the past, it has gained an increasing interest in the recent years. One of the many reasons is the increasing utilization of the syngas as a possible candidate for a new environmentally friendly source of energy [24]. The syngas, as a product of a gasification of carbon containing fuels including biomass, can be used to generate electricity by internal combustion gas turbines or engines, or can be transformed into chemicals. However, the syngas often contains various pollutants, especially stable PAHs–tars, that substantially disqualify it from further use [25]. Therefore, the syngas cleaning before its utilization is required. Besides gasification processes, the tars can also form in combustion processes of fossil fuels, although with lower amounts compared to gasification. However, their removal from exhaust gases is also important due to associated environmental issues [26].

Even though many published papers denoted as "plasma catalytic removal of model tar compound" can be found in the available literature, a majority of these papers actually deals with a removal of toluene (C_7H_8) as a model tar compound [27–32]. Nevertheless, the toluene represents "only" a monocyclic organic compound as it contains a single benzene ring. The real tars formed in the gasification and combustion processes are much more complex and usually consist of several benzene rings. Further, the stability of toluene is lower and its reactivity higher in comparison with e.g., naphthalene ($C_{10}H_8$), the simplest PAH containing two benzene rings. Indeed, toluene can be decomposed much more easily than naphthalene as a consequence of weaker C–C bonds in a toluene molecule due to the presence of methyl group—CH_3 [33]. Furthermore, decomposition of polycyclic tars with three or more benzene rings may lead to a formation of two-ring tars, just like naphthalene, that are often more stable than heavier polycyclic compounds [34]. For these reasons, a better candidate representing a group of tars is naphthalene rather than toluene.

A recent review paper written by Liu et al. focused on plasma catalytic removal of tars offers a very good overview of the field [18]. It summarizes the general knowledge, recent experiments, results, and findings of various authors. Indeed, several discharges have been employed and investigated in the naphthalene removal process by plasma catalysis: corona discharge [35,36], dielectric barrier discharge (DBD) [37–43], gliding arc discharge [44,45], and even plasma jet [46]. The discharges have been combined with several catalytic materials, most frequently with Ni-based catalysts (Ni/γAl$_2$O$_3$ [37,38,40,44], Ni/ZSM-5, Ni/SiO$_2$ [41] and Ni/Co-based catalyst [45]) due to their availability and selectivity towards formation of syngas constituents in catalytic processes [23]. However, their high activities are obtained only when operated at elevated temperatures (>780 °C) [47]. Besides Ni-based catalysts, other materials have also been investigated in a plasma catalytic process of naphthalene removal, including MnO_2 [35], γAl_2O_3 [36], Rh-LaCoO$_3$/Al$_2$O$_3$ [40], TiO$_2$/diatomite [43], and Pt/γAl_2O_3 [46].

The main objective of this work is to investigate tar removal by atmospheric pressure DBDs in combination with packing materials of various properties and catalytic activities. Naphthalene is selected as a model tar compound. The research follows our previous work [48], where naphthalene removal was investigated using TiO$_2$, Pt/γAl_2O_3, γAl_2O_3 and glass beads as packing materials. Here, we also study the effect of other materials including ZrO$_2$ and BaTiO$_3$. The used packing materials are characterized by distinct shape (spherical and cylindrical pellets and beads), size (3–5 mm in diameter, 3–8 mm in length), dielectric constant (5–4000), and specific surface area (SSA) (37–150 m^2/g). In comparison with the existing works of other authors, this work presents a comparative study on several packing materials of various properties and reports their plasma catalytic effects in naphthalene removal. The experiments are performed at a temperature of approx. 100 °C and a naphthalene initial concentration of approx. 3000 ppm, i.e., under conditions that are usually less favorable to achieve high removal efficiencies. Besides the effects of packing material properties on naphthalene removal efficiency, a formation of by-products of naphthalene decomposition is also studied. The by-products are analyzed by means of Fourier-transform infrared (FTIR) spectrometry and surface of packing materials by scanning electron microscope (SEM) analysis.

2. Results and Discussion

2.1. Discharge Characteristics

Prior to the experiments on naphthalene removal, we evaluated the discharge power of all DBD reactors by using the Lissajous figure method. The results showed that, at a given applied voltage, the reactors packed with $Pt/\gamma Al_2O_3$, γAl_2O_3, ZrO_2, TiO_2 (SSA of 70 m^2/g) and glass beads ($\varnothing$ 4 mm) had almost the same discharge power, whereas the power of plasma non-packed reactor as well as the reactor packed with $BaTiO_3$ significantly differed (see Figure 1a,b, where discharge power is expressed in terms of the specific input energy (SIE)). Note that uncertainty of the data points in Figure 1a,b is approx. 5–6%. Indeed, at the given applied voltage, the SIE of the plasma non-packed reactor was always higher than that of packed reactors (Figure 1a). On the contrary, the reactor packed with $BaTiO_3$ possessed the lowest discharge power in a range of the applied voltage of 11–14 kV (Figure 1a). Thus, a presence of the packing material inside the reactor generally decreases the discharge power under the same operating conditions. These findings are in agreement with the results of other authors, who also observed the same effect [49–51]. Inserting the packing material into a discharge gap has an influence on discharge mode [52,53], when the filamentary discharge mode typical for non-packed DBD reactors may change into surface discharge mode or into their combination [54]. Such change in discharge mode can also occur with the increasing of the applied voltage [55]. Furthermore, each pellet of packing material functions as a capacitor, so the charges can be trapped rather than being transferred across the gap, which also may affect the discharge current and its mode [49]. Thus, all these effects have an impact on the discharge properties including power consumption.

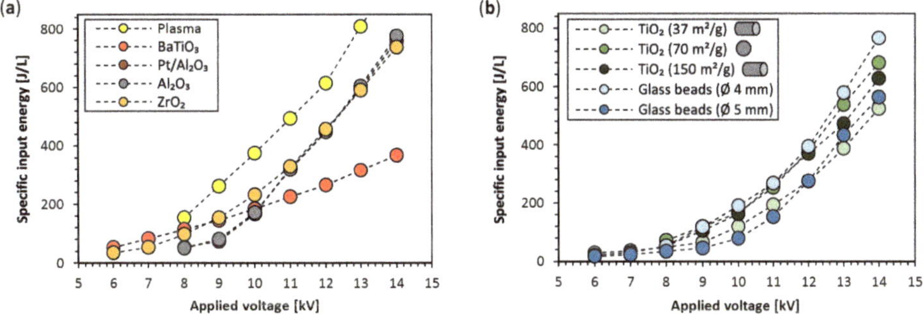

Figure 1. (**a**,**b**) Specific input energy as a function of applied voltage for various reactors (at 500 Hz).

In addition to the type of packing material, we also investigated the effects of its other properties (i.e., shape, size, SSA) on the discharge power. Packing materials of the two basic shapes were used. While packing materials including $BaTiO_3$, $Pt/\gamma Al_2O_3$, γAl_2O_3, TiO_2 (SSA of 70 m^2/g) and glass beads possess a spherical shape, ZrO_2 had a cylindrical pellet shape. However, a different shape of ZrO_2 in comparison with the other packing materials did not show any effect on discharge power (Figure 1a).

The effect of the SSA of packing material was also investigated in a case study of TiO_2. Three different SSAs were examined: TiO_2 with SSA of 70 m^2/g had a spherical shape ($\varnothing$ 3–4 mm), TiO_2 with SSA of 37 and 150 m^2/g had a cylindrical shape ($\varnothing$ 3–4 mm, length up to 7 mm). All parameters are indicated in the legend of Figure 1b for a clarity. When comparing TiO_2 with SSA of 37 and 150 m^2/g, the discharge power was found to be higher for TiO_2 with bigger SSA (Figure 1b). It can be explained by the fact that, besides discharges propagating in gas volume and on surfaces on the packing material, a plasma may be also formed inside the pores of packing material [56–58]. In such case, bigger SSA means higher pore volume and higher volume of generated plasma and, thus, higher discharge power. Nevertheless, at the highest amplitudes of the applied voltage (13 and 14 kV), spherical TiO_2 with moderate SSA of 70 m^2/g reached even higher discharge power than TiO_2 with bigger SSA of 150 m^2/g (Figure 1b). This result indicates that, at the highest tested amplitudes of the applied voltage, the effect of shape of packing material probably tends to dominate over the effect of SSA on discharge power.

Finally, the effect of size of packing material on discharge power was examined and also found to be significant. A discharge power of reactor packed with smaller glass beads (Ø 4 mm) was found to be higher than that of a reactor packed with larger glass beads (Ø 5 mm) (Figure 1b). It can be possibly explained by the fact that, when smaller beads are used in the same vessel, their total surface area is bigger, what can enhance a propagation of surface discharges in contrast to larger beads, where a shift from a full gap discharge to localized microdischarges occurs [52]. Thus, with smaller beads, the number of surface discharges and total volume of generated plasma is bigger, what leads to a higher discharge power. Another explanation of higher discharge power obtained with smaller beads is electric field enhancement in their contact points. Van Laer and Bogaerts reported a stronger electric field with smaller beads in contrast to bigger ones, what could eventually also lead to higher discharge power [52].

2.2. The Effect of Packing Material Type

Figure 2a,b presents naphthalene removal efficiency (NRE) and energy efficiency (EE) obtained with various plasma (non-packed) and plasma catalytic (packed) reactors. While the results obtained with TiO_2, $Pt/\gamma Al_2O_3$, γAl_2O_3, and glass beads were reported in our previous paper [48], the results obtained with $BaTiO_3$ and ZrO_2 are new and presented along with our previous results. Note that each experiment was usually performed 2–5 times. Data presented in Figure 2a,b (and also in Figure 3a,b) represent mean values of NRE and EE evaluated from all experiments. The error bars represent standard deviations of the data. In addition to plasma catalytic experiments, the effect of packing material alone (i.e., without a plasma) was also tested; however, no notable effect was observed on naphthalene decomposition and gaseous products formation.

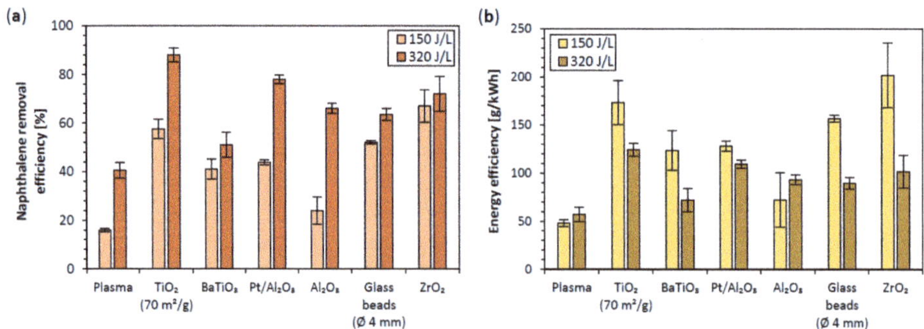

Figure 2. (**a**) Naphthalene removal efficiency and (**b**) energy efficiency for plasma and plasma catalytic reactors with various packing materials for the SIE of 150 and 320 J/L.

For the given SIE of 320 J/L, the NRE and the EE of all reactors followed a sequence: TiO_2 > $Pt/\gamma Al_2O_3$ > ZrO_2 > γAl_2O_3 > glass beads > $BaTiO_3$ > plasma only (Figure 2a,b). Thus, the highest NRE of approx. 88% for the SIE of 320 J/L was reached with plasma catalytic reactor packed with titanium dioxide TiO_2 (also called as titania) with a spherical pellet shape and SSA of 70 m^2/g. Furthermore, not only the highest NRE, but also the highest EE of 124 g/kWh was reached (for 320 J/L). High activity of TiO_2 in plasma catalytic removal of naphthalene was also reported by Wu et al. [43], who obtained a similar NRE of 88%, however, at lower naphthalene initial concentration of 60 ppm. High NRE may be probably attributed to its photocatalytic activity initiated by the absorption of UV radiation emitted by plasma [59,60] or, more likely, by interaction with plasma-produced highly energetic electrons [8,61,62]. When a reactor packed with TiO_2 without a plasma was tested, no effect on naphthalene decomposition was observed probably due to the absence of UV radiation and energetic electrons that are crucial for its activation [63].

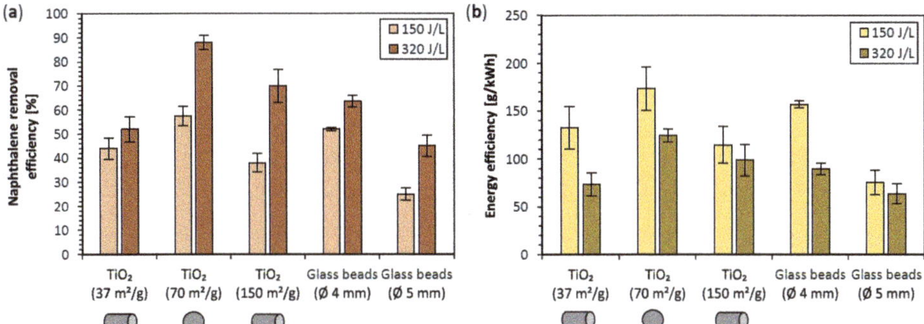

Figure 3. (**a**) Naphthalene removal efficiency and (**b**) energy efficiency for plasma catalytic reactors packed with TiO_2 of various SSAs and shapes and with glass beads of various diameters for the same SIE of 320 J/L.

The second highest NRE and EE of 78% and 110 g/kWh for the SIE of 320 J/L was obtained with a $Pt/\gamma Al_2O_3$ catalyst (Figure 2a,b). A discussion related to the results obtained with $Pt/\gamma Al_2O_3$ can be found in our previous paper [48]. The $Pt/\gamma Al_2O_3$ catalyst was also tested by Yuan et al. in combination with radio-frequency-powered plasma jet in ambient air assisted by Ar mixture for naphthalene and n-butanol removal [46]. The authors reported a superior increase of the NRE when a catalyst was used (approx. 99%) compared to a reactor without it (30%). Various other Pt-based catalysts have also been extensively investigated on catalytic naphthalene removal and their high activity was commonly reported [64–66]. In addition to high catalytic activity, high CO_2 selectivity was also observed [64,67]. These works showed that Pt-based catalysts are efficient not only for catalytic VOC removal as is generally well-known [68,69], but are also very active for catalytic decomposition of more complex molecules including PAHs. However, Ndifor et al. and Shie et al. reported catalytic activity of $Pt/\gamma Al_2O_3$ towards naphthalene oxidation only above 150 °C [64,65], i.e., at temperatures higher than in our experiment. Our results showed that $Pt/\gamma Al_2O_3$ may also be efficient at a lower temperature of 100 °C, but only when plasma catalysis is employed.

Another tested metal oxide catalyst was zirconium dioxide ZrO_2 (zirconia). Although for 320 J/L the zirconia reached the third highest NRE, for lower SIE of 150 J/L, it reached the highest NRE and EE among all tested packing materials of approx. 67% and 202 g/kWh, respectively. The zirconia is widely used as a catalyst support, but it also provides an intrinsic catalytic activity [70,71]. Moreover, a very promising catalytic activity of zirconia for high-temperature oxidation of tars was observed by several authors [72–74]. According to our results, we suppose that the zirconia, like $Pt/\gamma Al_2O_3$, may manifest a high catalytic activity even at lower temperatures when activated by plasma. Another reason for a relatively high NRE obtained with zirconia can be, like titania, associated with its photocatalytic activity, although its band gap is much wider (5.5–7 eV), than that of titania (3–3.2 eV) [75,76]. As the intensity of the UV radiation generated by air atmospheric pressure discharges necessary for zirconia photocatalytic activation (e.g., below 225 nm) is typically very low [77], we assume that the catalyst activation can only be initiated by plasma-produced highly energetic electrons [8,61,62].

The NRE obtained with γAl_2O_3 (alumina) and glass beads were almost identical: 66 and 64% for the SIE of 320 J/L. More details can be found in [48]. Plasma catalytic removal of naphthalene by using pulsed corona discharge combined with γAl_2O_3 was also investigated by Nair [36]. He reported an increase of the NRE (>95%) and a reduction of energy demand when plasma catalysis was employed in contrast to the case of plasma alone. He worked at a temperature of 300 °C and at an initial naphthalene concentration of 500–700 ppm. Although γAl_2O_3 is the most commonly used bulk material for the supported catalysts, its properties may also play an important role in enhancing the catalytic performance of supported catalysts [78,79]. In addition, γAl_2O_3 has an ability to adsorb organic compounds [80], which may increase the overall efficiency of their removal and even to enhance CO oxidation [81]. Moreover, Roland et al. suggested that γAl_2O_3 is able to enhance a decomposition of ozone O_3 leading to a formation of O radicals, what in turn may enhance plasma catalytic VOC oxidation [82]. This effect was also proposed by Wu et al. for observed improvement in naphthalene removal over TiO_2/diatomite catalyst, as the diatomite is composed of approx. 16% of γAl_2O_3 [43]. With regard to the effect of glass beads as a packing material, their utilization in naphthalene removal was also studied by Hübner et al. [39] and Redolfi et al. [42]. They obtained an NRE of 60 and 95%, however, at higher operating temperatures of 350 and 250 °C and at significantly lower naphthalene initial concentrations of 90 and 100 ppm, respectively. Even though the glass beads represent a dielectric material without any specific catalytic activity, their role in improvement of the NRE probably lies in affecting the discharge characteristics, quality, and distribution [83].

Finally, the reactor packed with $BaTiO_3$, a high dielectric constant material (~4000), was also tested. A purpose for introducing of ferroelectric packing materials in a plasma reactor is generally to amplify the electric field. Because of this, a distribution of electrons is shifted towards higher energies and, thus, a generation of high-energy species in the discharge zone is established [84]. Although inserting the $BaTiO_3$ resulted in a significant change of characteristics of discharge current pulses (not showed), the chemical effects of $BaTiO_3$ reactor were found to be relatively weak. The NRE obtained with $BaTiO_3$ was the lowest among all plasma catalytic packed reactors: for the SIE of 320 J/L, the NRE and EE were 51% and 72 g/kWh, respectively (Figure 2a,b). The high dielectric constant packing materials constrain a discharge distribution predominantly to the contact points of the pellets, regardless of their contribution to electric field enhancement [53]. The discharge exhibits a filamentary microdischarge mode without surface discharges propagating along the packing material. This limits a catalyst surface area exposed to plasma resulting in a low catalytic activity. Dielectric constant of the other packing materials tested in our study is considerably lower (glass beads~5, γAl_2O_3~9, ZrO_2~20, TiO_2~85, $Pt/\gamma Al_2O_3$~100), what is in favor of the formation of surface discharges in these reactors (indirect evidence of their presence is presented in Section 2.4.1). Based on our results, we can conclude that a key factor for obtaining a substantial NRE is a presence of surface discharges propagating along the packing material. Firstly, they distribute a plasma more uniformly and intensively within a reactor and, secondly, they activate a larger area of the catalyst. The importance of surface discharges for attaining good catalytic activity in plasma catalysis was also confirmed by Kim et al. [85] and Veerapandian et al. [16]. On the other hand, high dielectric constant materials may find a possible utilization in applications where high-energy electrons are required in order to obtain good efficiencies (i.e., conversion of CO_2) [19,86,87].

2.3. The Effect of Packing Material Properties (SSA, Shape, and Size)

The effect of packing material properties (SSA, shape, and size) on NRE and EE was also studied and evaluated. Although the effect of SSA on plasma physical characteristics as well as SIE was relatively small (Figure 1b), it was expected to have much stronger impact on chemical processes on the surface of the packing material. An increase of the SSA leads to enhancement of adsorption of target compounds on the material surface that may further support their oxidation and formation of by-products [16]. This effect was also observed for naphthalene when it was adsorbed on materials of various SSAs [88,89]. Adsorption of target compounds may also be influenced by other textural properties of packing materials (e.g., pore size and volume). Furthermore, plasma reactive species can diffuse into catalyst pores, which may lead to an enhanced catalytic activity [57]. On the other hand, a variation of shape and size of packing materials may alter and enhance the electric field, and consequently plasma properties. Plasma formation and distribution inside the packed reactor may be affected as well, what can also eventually lead to enhanced chemical effects under plasma catalysis [16,52,86].

In the previous Section 2.2., we presented the results of the effects of various types of packing materials on the NRE. A presented sequence of the used materials ordered with respect to the achieved NRE is valid, although the effect of SSA was not explicitly considered despite it possibly affects the overall results. The SSA of a material is one of the most critical parameters for conventional catalysis. In order to accurately assess the effects of various types of packing materials, they should possess a similar SSA. Although the SSA of the used packing materials was not always known, our results showed that the role of SSA in plasma catalysis is not decisive as it is in conventional catalysis. More specifically, γAl_2O_3 and glass beads are characterized by very distinct SSAs (typically 250–300 m^2/g [85] and <0.001 m^2/g, respectively). Despite the huge difference in the SSA of the two materials, our results indicate that for the same SIE of 320 J/L, the NRE for γAl_2O_3 and glass beads is almost the same (66% and 64%, respectively). Therefore, we suppose that, in plasma catalysis, in addition to SSA, other parameters may play more important roles. Nevertheless, in this study, the effect of SSA along with the effect of shape of packing material was investigated in a case study of TiO_2.

Figure 3a,b presents the NRE and EE for reactors packed with TiO_2 catalysts of various SSAs and shapes and glass beads of various sizes. With an increase of SSA (37 vs. 150 m^2/g), we observed an increase of NRE and EE for the same SIE (52 vs. 70% and 74 vs. 99 g/kWh for 320 J/L, respectively). It can be probably explained by enhanced adsorption processes on the catalyst surface and, thus, more efficient naphthalene decomposition. A change of packing material shape (spherical vs. cylindrical) was found to have an even stronger effect than a change in the SSA with respect to NRE and EE, as a spherical TiO_2 catalyst of 70 m^2/g showed higher NRE and EE (88% and 124 g/kWh for 320 J/L, respectively) than cylindrical TiO_2 with a bigger SSA of 150 m^2/g (70% and 99 g/kWh for 320 J/L, respectively). This result can be explained by the fact that the shape of the packing material determines the size of voids/spaces between the pellets, and thus discharge properties and plasma volume (this effect is also discussed later in Section 2.4.1). The result also demonstrates one of the main differences between plasma catalysis and conventional catalysis, in which SSA (or textural properties of packing materials in general) plays a crucial and often a decisive role. In plasma catalysis, not only textural properties of packing materials, but also properties of plasma as well as complex interactions between them determine the overall effect.

Finally, our results also revealed a positive effect of smaller glass beads (Ø 4 mm) when compared to larger glass beads (Ø 5 mm), as the NRE and EE were higher with smaller beads for both tested SIE values (for 320 J/L, 64 vs. 45% and 90 vs. 64 g/kWh, respectively). It can possibly be explained by a bigger total surface area of smaller beads that consequently leads to an enhancement of surface discharges propagation along the surface of beads. In the case of larger beads, a shift from full gap discharge to localized microdischarges occurs [52]. For this reason, an enhancement in surface discharges, as we stated earlier, could finally lead to higher removal efficiency. However, Van Laer and Bogaerts as well as Butterworth et al. observed more complicated relationship between removal and energy efficiencies and packing material size, especially when the size of packing material becomes very small [86,90,91]. Van Laer and Bogaerts reported higher removal and energy efficiency with a packing material of diameter bigger than 1.5 mm, than for a non-packed reactor (for the discharge power of 60 W). However, when using packing of diameter smaller than 1.5 mm, removal and energy efficiency were lower than for a non-packed reactor probably due to very short gas residence time in a reactor [90,91]. Furthermore, Butterworth et al. reported that the use of smaller packing material sizes (0.18–2 mm in diameter) can lead to either an increase or a decrease of removal efficiency depending on various other conditions and parameters (e.g., gas mixture) [86]. Therefore, the removal efficiency is not only given by a balance between electric field enhancement and gas residence time given by the used packing material size, but it also depends on other parameters as well as on the type of packing material [92].

2.4. Analysis of Naphthalene Decomposition By-Products

The evaluation of NRE and EE was supplemented by the analysis of naphthalene decomposition by-products. In principal, an exact composition and yield of by-products were strongly dependent on several parameters, such as applied voltage, discharge power, reactor type, and properties of packing material. The by-products were found in the gas phase as well as in the solid phase as deposits on the reactor walls, inside the gas lines, on windows of the gas cell and surface of packing materials. The two diagnostic methods used for analysis of by-products were scanning electron microscopy (SEM) equipped with energy-dispersive X-ray (EDX) spectroscopy and infrared absorption spectroscopy FTIR. The optical microscope was also used for taking the photographs of packing materials surface.

2.4.1. SEM Analysis

A surface analysis of both fresh and used packing materials (i.e., before and after the experiment, respectively) was carried out by means of the SEM. For the analysis, both secondary and backscattered electrons were utilized to obtain surface morphology and to distinguish between areas with different chemical composition, respectively. As electrical resistivity of fresh and used packing materials was quite different, the amplitude of accelerating voltage of scanning electrons for each material was adjusted individually (1.5–20 kV) in order to avoid overcharging of the samples, which could negatively affect an image sharpness. Besides the SEM analysis, we also performed the EDX analysis in order to determine an elemental composition of analyzed surface.

For the fresh packing materials, the SEM analysis allowed us to determine a surface morphology of the pellets, since it consists of pores, granules, and structures of various shapes and sizes. For used packing materials, we identified a shape and a size of solid deposits formed on a surface of the pellets. In addition, we also performed a cross-sectional analysis of used materials after we cut the pellets in half. This allowed us to estimate a thickness of solid deposits accumulated on the surface and also the depth of their penetration inside the pores of packing materials. Table 1 summarizes SEM images and optical microscope images of selected fresh and used materials. Note that various magnifications are used in SEM images, as solid deposits on surface of various packing materials possessed a different size.

Table 1. SEM and optical microscope images of fresh, i.e., before experiment (left column), and used, i.e., after the experiment (right column), selected packing materials (magnification in optical microscope images: 160×).

Fresh Packing Materials (Before Experiment)	Used Packing Materials (After Experiment)
TiO_2 (70 m²/g)	

Table 1. *Cont.*

Fresh Packing Materials (Before Experiment)	Used Packing Materials (After Experiment)

TiO_2 (37 m^2/g)

$BaTiO_3$

$Pt/\gamma Al_2O_3$

ZrO_2

At first sight, we can see significant differences between photographs of packing materials before and after experiments. The surface of all materials after the experiments acquired a brownish color (Table 1, right column). The change in surface color of materials is attributed to by-products of naphthalene decomposition that formed solid deposits on the surface as a result of the incomplete oxidation of naphthalene. The EDX analysis showed an elemental composition of the deposits and, as was expected, they were mostly composed of carbon and oxygen. The SEM analysis revealed that the solid deposits were composed of round particles with a diameter in the range of 1–5 µm. These particles also showed strong agglomerating tendencies. A very similar shape of solid carbon deposits resulting from naphthalene decomposition was found by Wang et al.; however, they reported a size of particles of two orders of magnitude smaller than we observed (40–60 nm in diameter) [93]. Such a significant difference in size of carbon particles can be probably explained by different working conditions (temperature of 200 °C, N_2 + H_2O as a carrier gas).

Figure 4a shows an SEM image of a cross-sectional view of TiO_2 surface. Based on this image, it is possible to estimate a thickness of solid carbon deposits (being up to 10 µm) accumulated on the TiO_2 surface for 3.5 h of experiment. Figure 4b–f shows detailed SEM images of aggregates of solid deposits on the surface of various packing materials and confirms that round-shape particles of carbon deposits were observed regardless of the packing material type.

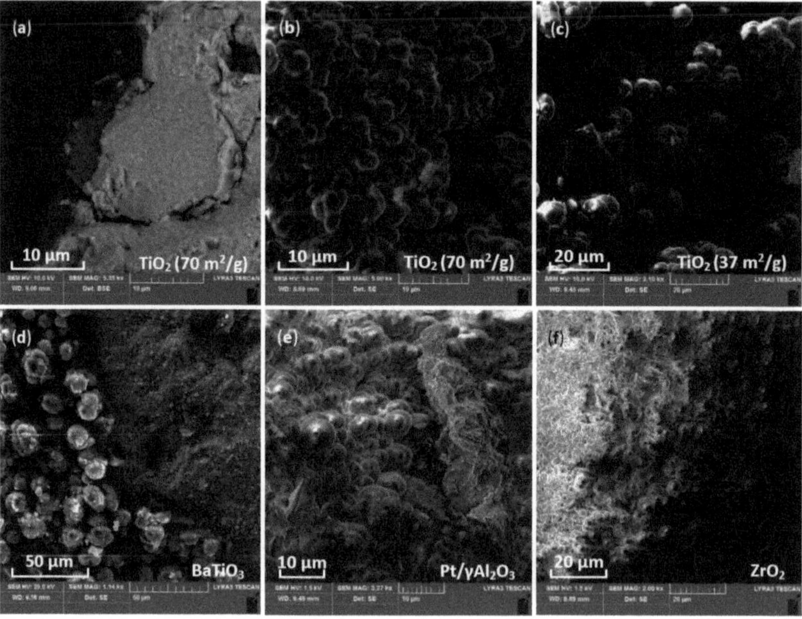

Figure 4. Detailed SEM images of (**a**) cross-sectional view of TiO_2 surface and (**b–f**) solid carbon deposits on surface of various packing materials. Note that different magnifications are used.

The packing materials of spherical shape (TiO_2 of 70 m²/g, Pt/γAl_2O_3 and γAl_2O_3) were almost uniformly covered with solid deposits. On the contrary, cylindrical materials (TiO_2 of 37 and 150 m²/g and ZrO_2) were characterized by nonuniform distribution of deposits (see the images in Table 1). The most specific solid deposits were found on the $BaTiO_3$ surface, where they agglomerated in circular patterns with a diameter of approx. 1 mm randomly distributed on the surface (Figure 5a,b). The EDX analysis showed that the circular deposits consisted of carbon and oxygen, while the surrounding was composed of Ba and Ti (Figure 5c,d). The explanation for a formation of such specific circular patterns of deposits can be given as follows: the reactor packed with $BaTiO_3$ (i.e., a high dielectric constant material) is governed by filamentary microdischarge mode [52,53,94]. These microdischarges are formed only near the contact points of adjacent pellets where an electric field is extremely strong and where the reactive species are predominantly formed [53]. In the filamentary microdischarge mode, formation of surface discharges propagating along the surface of packing material is completely suppressed, what in turn leads to limited surface area of packing material exposed to generated plasma [6]. Moreover, production of reactive species along the surface of packing material is suppressed too. Therefore, concentration of reactive species far from the regions of strong electric field significantly decreases. Thus, we hypothesize that naphthalene decomposition process took place only in a close vicinity of filamentary microdischarges (i.e., near the contact points of pellets). Hence, a formation of by-products was also confined in the same regions where a strength of electric field and concentration of reactive species were sufficiently high resulting in circular patterns of solid deposits. Note that most of these circular patterns have a void in their center. We assume that the voids correspond to the places where the adjacent pellets touched each other. Therefore, the solid deposits did not form here.

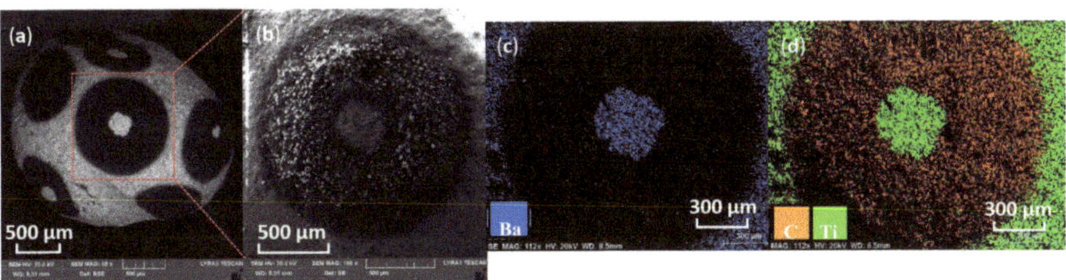

Figure 5. (**a**,**b**) SEM images and (**c**,**d**) EDX analysis of circular patterns of solid deposits on $BaTiO_3$ surface.

In addition to $BaTiO_3$, a few circular patterns of solid deposits were also found on the surfaces of cylindrical pellets (ZrO_2, TiO_2 of 37 m^2/g) (Figure 6), whereas on spherical pellets we did not observe them at all. Based on this observation, we further hypothesize that, in the case of spherical pellets of low and moderate dielectric constant, the surface discharges were much more intense than in the case of cylindrical pellets, and resulted in almost uniform distribution of solid by-products along the entire spherical surface. On the other hand, for cylindrical pellets, some circular patterns of solid deposits were found probably as a result of intense filamentary microdischarges rather than surface discharges. This implies that, in addition to a dielectric constant, a packing material shape may also partially affect a discharge mode. The cylindrical pellets have sharper edges that may significantly contribute to local electric field enhancement and, thus, to formation of multiple filamentary microdischarges in contrast to spherical pellets [95]. Thus, the generated plasma is in contact with a limited surface area of packing material, what in turn leads to spatially non-uniform naphthalene decomposition and distribution of by-products along the entire surface (the same, but even stronger effect was observed with $BaTiO_3$ as mentioned above). Consequently, a confinement of plasma formation to a vicinity of contact points of packing material in contrast to plasma formation and propagation along the surface of packing material (cylindrical vs. spherical pellet, respectively) can possibly explain somewhat counterintuitive results, when the reactor packed with a cylindrical TiO_2 of bigger SSA reached smaller NRE than a spherical TiO_2 of smaller SSA (Figure 3a).

Figure 6. Photographs of surface of various packing materials with circular patterns of solid deposits (magnification 250×).

2.4.2. FTIR Analysis

The FTIR analysis was conducted for the identification of gaseous and solid by-products of naphthalene decomposition. The gaseous and solid by-products were analyzed individually. Spectra of solid by-products (deposits on windows of the gas cell) were obtained after the experiment when discharge and naphthalene flow were turned off, but solid deposits remained on windows of the gas cell. The spectra of gaseous products were obtained indirectly. Firstly, the combined spectra of gaseous products together with solid by-products were recorded during the experiment. Subsequently, the spectrum of solid by-products was subtracted from the combined spectrum of gaseous and solid by-products to obtain a spectrum of only gaseous products. Thereby, the subtracted spectrum of gaseous products showed narrow absorption bands, while spectra of solid by-products were characterized by intense and very broad bands. Figure 7 shows typical FTIR spectra of gaseous and solid by-products (deposits on the gas cell windows), and the most relevant absorption bands are marked.

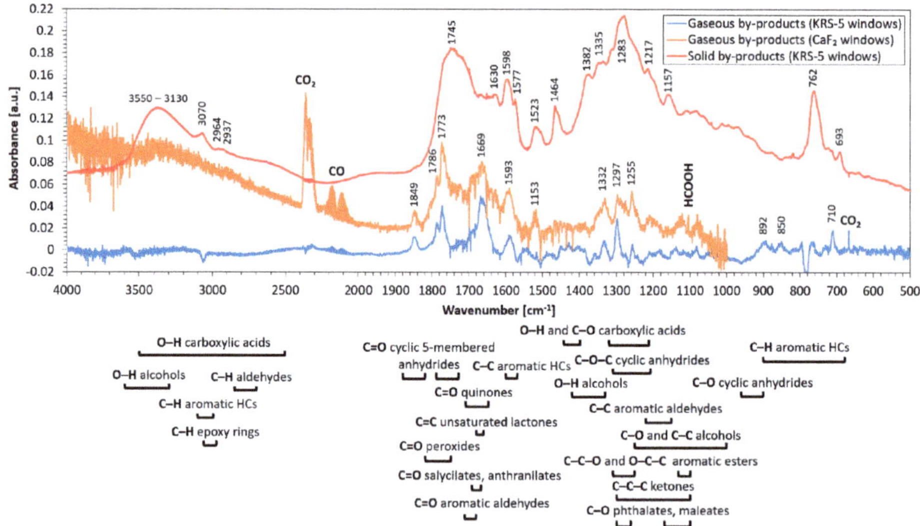

Figure 7. Infrared absorption spectra of gaseous (blue and orange line) and solid (red line) by-products of naphthalene decomposition (HCs = hydrocarbons).

The main gaseous products of naphthalene decomposition identified in the FTIR spectra were carbon monoxide CO, carbon dioxide CO_2, water H_2O and formic acid HCOOH, along with the nitrous oxide N_2O and ozone O_3 formed by the reactions in the carrier gas. The concentrations (i.e., production) of two main gaseous products CO and CO_2 depended on reactor type, discharge power and carrier gas. In this work, we used only synthetic air as a carrier gas, while, in our previous work, we also tested the effect of nitrogen and oxygen. In nitrogen, no gaseous products were found, while, in ambient air and oxygen, CO and CO_2 dominated [48]. Among all reactors, the highest CO_2 concentration was obtained with Pt/γAl$_2$O$_3$ (approx. 1140 ppm), while CO concentration was only approx. 180 ppm for 320 J/L. On the contrary, plasma catalytic TiO$_2$ reactor owned lower CO_2 production (approx. 610 ppm) and higher CO production (approx. 290 ppm) for the same SIE of 320 J/L. It indicates that, although higher NRE was obtained with TiO$_2$, more efficient oxidation of naphthalene and formation of gaseous products was observed with Pt/γAl$_2$O$_3$. In addition to CO and CO_2, a maximum concentration of HCOOH was found to be approx. 80 ppm. When considering only CO, CO_2, and HCOOH concentrations, the obtained carbon balance (CB) for all reactors was found to be in a range of 2–4%. More specifically, a plasma non-packed reactor reached CB of approx. 2%, while the reactor packed with $Pt/\gamma Al_2O_3$ of approx. 4%. These CB values are indeed very low and significantly underestimated, because other more complex compounds identified among the by-products (discussed below) have not been taken into account, as we identified them only qualitatively and not quantitatively. Low values of CB further imply a low proportion of CO, CO_2, and HCOOH, and, on the contrary, a very high proportion of other more complex compounds among the by-products. This represents a weakness of our results when compared to the results of other authors. However, an improvement of CB is a scheduled part of our future research.

Besides the main gaseous products of naphthalene decomposition, other more complex gaseous as well as solid compounds were also found in the spectra, although their analysis was not trivial. We reviewed available atlases and databases of FTIR spectra in order to find characteristic organic functional groups and then assigned them to absorption bands in the obtained spectra [96–98]. This approach also allowed for identification of individual compounds in the spectra.

A detailed FTIR spectra analysis of gaseous products revealed that the most intense absorption bands belong to the carbonyl functional group C=O, which is characteristic for a group of compounds including ketones, aldehydes, carboxylic acids, esters, and carbonates. Among the possible compounds, the carboxylic acid anhydrides seem to have the best match with the spectra. As anhydrides contain two carbonyl groups bonded by the same oxygen atom, they show two characteristic C=O absorption bands. The best match was found for cyclic 5-membered carboxylic acid anhydrides that possess unique feature in contrast to straight chain anhydrides: the band at a lower wavenumber is more intense than the band at a higher wavenumber. In general, cyclic anhydrides show two carbonyl bands in the range of 1870–1845 cm^{-1} and 1800–1775 cm^{-1} (Figure 7) [96]. Out of the anhydrides, the best matches were obtained for phthalic anhydride (1845 and 1775 cm^{-1}), maleic anhydride (1848 and 1790 cm^{-1}), and naphthalene-1:2-dicarboxylic anhydride (1848–1845 and 1783–1779 cm^{-1}) [98].

At lower wavenumbers (1720–1620 cm^{-1}), another intense C=O band can be found. The analysis showed a good match with quinones, the group of ketones, which are compounds derived from aromatics by substitution of even number of C–H groups by C=O groups. When two C=O groups are attached to one ring, the quinones absorb in a range of 1690–1660 cm^{-1} (Figure 7). Particularly, the 1,2- and 1,4-benzoquinone absorb around 1669 and 1667 cm^{-1}, while 1,2- and 1,4-naphthoquinone around 1678 and 1675 cm^{-1}, respectively [98].

For cyclic anhydrides, two other absorption bands of C–O and C–C are also relatively intense and can be found at 1300–1180 cm^{-1} and 960–880 cm^{-1} (Figure 7) [96,99]. In the case of quinones, absorption in a range of 1350–1200 cm^{-1} was reported [98]. Furthermore, characteristic bands of aromatic compounds are also present in the spectra. They can be attributed to naphthalene itself or rather to naphthalene-derived compounds. In general, the characteristic C–H absorption bands of aromatic rings occur at 3100–3000 cm^{-1} and 900–675 cm^{-1}, while C–C bands of medium intensity can be found at 1600–1585 cm^{-1} and 1500–1400 cm^{-1} (Figure 7) [96,99].

The spectra of solid by-products were analyzed in the same way as gaseous products and several matches for the same compounds were found. However, the analysis was much more complicated due to very broad absorption bands which are typical for solid compounds and also as a result of individual bands overlapping. One of the most striking differences between spectra of gaseous and solid by-products is the presence of very broad band at high wavenumbers in the range of 3600–2700 cm^{-1} (Figure 7). The band arises from absorption of hydroxyl O–H group (3500–2500 cm^{-1}) usually attributed to alcohols but here more likely to carboxylic acids [96]. In addition, the carboxylic acids also contain bands as follows: C=O (1730–1680 cm^{-1}), C–O (1320–1210 cm^{-1}) and O–H (1440–1395 cm^{-1} and 960–900 cm^{-1}) (Figure 7). A good match was also found with salicylic acid (C=O band at 1655 cm^{-1}) and benzoic acid (C=O band at 1685 cm^{-1}) [98].

Finally, gaseous as well as solid compounds identified in the FTIR spectra based on detailed analysis of functional groups were confronted with the compounds reported as by-products in the works of other authors related to naphthalene decomposition by plasma and plasma catalysis [24,33,35,36,40,44,93,100–104]. The comparison showed a good match and supported our findings. Thus, we can conclude that, among the complex gaseous products, phthalic anhydride, maleic anhydride, 1,4- and 1,2-naphthoquinone, 1,4-benzoquinone, 1,8-naphthalic anhydride and 3,4-naphthalene-1,2-dicarboxylic anhydride were positively identified. In a solid phase, salicylic acid, phthalide, 1-naphthalenecarboxylic acid, and 1,8-naphthalic anhydride were positively identified. The compounds found in the spectra represent organic gaseous and solid intermediates of naphthalene decomposition and indicate its incomplete oxidation to the desired products of CO_2 and H_2O.

3. Materials and Methods

The experimental setup is depicted in Figure 8. The NTP was generated by atmospheric pressure DBD reactors of cylindrical geometry. The reactors consisted of quartz glass tube (15 mm in diameter, 10 cm in length) and tungsten wire (0.2 mm in diameter) placed in the axis of the tube and aluminum foil wrapped around the tube. The wire served as a high voltage electrode, while the aluminum foil was grounded. The plasma (non-packed) or plasma catalytic (packed) reactors were powered by AC high voltage power supply consisting of a function generator (GwInstek SFG-1013, Taipei, Taiwan), signal amplifier (Omnitronic PAP-350, Waldbüttelbrunn, Germany), and high voltage transformer. The main set of experiments was performed at a frequency of 200 and 500 Hz and for the SIE of 150 and 320 J/L, respectively. Waveforms of the applied voltage and the discharge current were monitored by a high voltage probe (Tektronix P6015A, Berkshire, UK) and current probe (Pearson Electronics 2877, Palo Alto, CA, USA), respectively, both connected to a digital oscilloscope (Tektronix TBS2000, Berkshire, United Kingdom). The power consumption of the reactors was evaluated using the Lissajous figure method with an 82 nF capacitor and a voltage probe (Tektronix P2220, Berkshire, UK). The capacitances of the used reactors were in a range of approx. 1–10 pF.

Synthetic air (purity 5.0) supplied from the pressure tank was used as a carrier gas, and its flow rate was controlled by mass flow controllers (MFC) (Bronkhorst El-Flow Prestige FG-201CV, Ruurlo, Netherlands). The experimental system including the DBD reactors and gas lines was heated using an electric oven and ribbon heaters to a temperature of 100 °C. Naphthalene (CentralChem, Banská Bystrica, Slovakia) was used as a model tar compound. Two gas lines were used and led to the oven. The air from one line was enriched with naphthalene vapors and then mixed with the air from the second line in order to obtain a desired input naphthalene concentration (approx. 3000 ppm), while total air flow rate was set to 0.5 L/min. The gaseous and solid by-products of naphthalene decomposition were analyzed by means of a Fourier-transform infrared (FTIR) spectrometer (Shimadzu IR-Affinity 1S, Kyoto, Japan) using a gas cell with an optical path length of 10 cm equipped with either CaF_2 or KRS-5 windows. Main gaseous products (CO, CO_2, and HCOOH) were identified in the FTIR spectra, and their concentrations were evaluated using absorption bands as follows: CO at 2180 cm^{-1}, CO_2 at 2360 cm^{-1} and HCOOH at 1105 cm^{-1}. The solid by-products (deposits) founded on windows of the gas cell were analyzed along with gaseous products. A surface analysis of the fresh and used packing materials was performed by scanning electron microscope (SEM) (Tescan Lyra3, Brno–Kohoutovice, Czech Republic) equipped with an energy-dispersive X-ray spectroscopy (EDX) detector (Bruker 129 eV, Billerica, MA, USA). In addition, the optical stereo microscope (Leica Wild M3C, Wetzlar, Germany) was used for taking the photographs of packing materials surface.

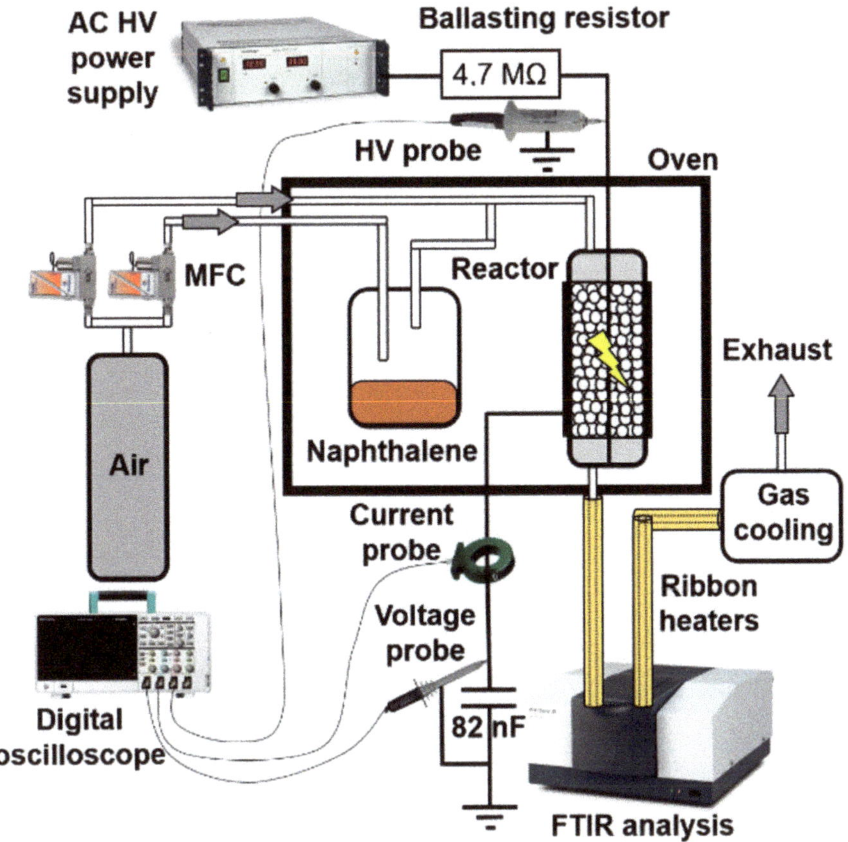

Figure 8. Experimental setup.

The plasma catalytic reactors were filled with various packing materials including catalysts (TiO_2, $Pt/\gamma Al_2O_3$, ZrO_2) and other dielectric materials ($BaTiO_3$, γAl_2O_3, glass beads). The packing materials were characterized by distinct properties (dielectric constant, shape, size, and specific surface area) summarized in Table 2.

A typical procedure of an individual experiment was comprised of several steps. Firstly, an air with naphthalene vapors was allowed to flow into a reactor and at the same time the discharge was turned on and the SIE was set to 150 J/L. The decomposition of naphthalene induced by the discharge resulted in an increase of CO and CO_2 concentrations, whereas the naphthalene concentration (evaluated at 781 and 3067 cm^{-1}) also gradually increased until it fully stabilized. Once all measured concentrations stabilized, the SIE was increased and set to 320 J/L. Due to an increase of the SIE, naphthalene concentration gradually decreased, and CO and CO_2 concentrations correspondingly increased as more naphthalene was converted to CO and CO_2. When all measured concentrations stabilized again, the discharge was turned off, while air with naphthalene vapors was still allowed to flow into a reactor. The CO and CO_2 concentrations dropped down to zero, while naphthalene concentration gradually increased until it reached the input concentration. Finally, the naphthalene was turned off and only air was allowed to flow. The experiment was finished once the naphthalene concentration dropped down to zero. The entire experiment typically lasted for 3–7 h. The time depended on the used catalyst and its properties that determined waiting times of the individual steps necessary to reach stabilized concentrations.

Table 2. Properties (dielectric constant, shape, size, specific surface area (SSA) and producer) of used packing materials (N/A = not available).

	TiO_2	$BaTiO_3$	$Pt/\gamma Al_2O_3$	γAl_2O_3	Glass beads	ZrO_2
Dielectric constant	20–100 (~85) [52,105]	400–10000 (~4000) [86,106]	~100 [107]	8–11 (~9) [52,85]	~5 [52]	12–25 [19,52,108]
Shape	Spherical, cylindrical	Spherical/ cylindrical	Spherical	Spherical	Spherical	Cylindrical
Size [mm] (Ø = diameter; L = length)	Ø 3–4 (sphere) Ø 3–4, L 3–7 (cylinder)	Ø 3–5	Ø 3–4	Ø 3–4	Ø 4, Ø 5	Ø 3, L 4–8
SSA [m²/g]	37 (cylinder), 70 (sphere), 150 (cylinder)	N/A	N/A	typically 250–300 [85]	<0.001	90
Producer	Sakai Chemicals Co./Abcr GmbH (Osaka, Japan/Karlsruhe, Germany)	Fuji Titanium Co. (Osaka, Japan)	*unknown*	Sumitomo Chemical Co. (Tokyo, Japan)	*unknown*	Abcr GmbH (Karlsruhe, Germany)

The following variables were defined in order to evaluate a performance of DBD reactors:

- *Specific input energy* (SIE) (i.e., energy density or discharge energy per gas volume):

$$\text{SIE (J/L)} = \frac{P}{Q}\,, \tag{1}$$

where P and Q represent power consumption of the discharge and total gas flow rate, respectively.

- *Naphthalene removal efficiency (NRE):*

$$\text{NRE (\%)} = \left(1 - \frac{c}{c_0}\right) * 100\,, \tag{2}$$

where c_0 and c represent the input and output concentrations of naphthalene, respectively.

- *Energy efficiency (EE)* (i.e., the amount of removed naphthalene per specific input energy):

$$\text{EE (g/kWh)} = \frac{\text{NRE} * c_0}{\text{SIE}}\,, \tag{3}$$

- *Carbon balance (CB)* (i.e., the sum of quantified products per amount of removed carbon):

$$\text{CB (\%)} = \frac{\sum c_{\text{products}}}{n * (c_0 - c)} * 100\,, \tag{4}$$

where $\sum c_{\text{products}}$ represents the sum of concentrations of all products, while n is the number of carbon atoms in molecule of naphthalene (10).

4. Conclusions

In this paper, we investigated the removal of tars by plasma catalysis with naphthalene ($C_{10}H_8$) as a model tar compound. The NTP was generated by atmospheric pressure DBD reactors and in combination with various packing materials including catalysts (TiO_2, $Pt/\gamma Al_2O_3$, ZrO_2) and other dielectric materials ($BaTiO_3$, γAl_2O_3, glass beads). The packing materials were characterized by a distinct shape (spherical and cylindrical pellets and beads), size (3–5 mm in diameter, 3–8 mm in length), dielectric constant (5–4000), and specific surface area (SSA) (37–150 m^2/g). The effects of packing material types and their properties on plasma characteristics, naphthalene removal, and formation of by-products were studied. For the given SIE of 320 J/L, the naphthalene removal efficiency (NRE) of all reactors followed a sequence: TiO_2 (88%) > $Pt/\gamma Al_2O_3$ (78%) > ZrO_2 (72%) > γAl_2O_3 (66%) > glass beads (64%) > $BaTiO_3$ (51%) > plasma only (41%). An improvement of the NRE was observed with increasing SSA of the catalyst; however, the shape of catalyst pellets was found to be even more important: the NRE obtained with cylindrical TiO_2 with SSA of 37 and 150 m^2/g were 52 and 70%, respectively, while spherical TiO_2 with moderate SSA of 70 m^2/g achieved even higher NRE of 88% (for 320 J/L). In addition, improvement of the NRE was also found with smaller glass beads (Ø 4 mm) when compared to larger glass beads (Ø 5 mm) (for 320 J/L, 64 vs. 45%, respectively).

The by-products formed by naphthalene decomposition were found in the gas phase as well as in the solid phase as deposits on the windows of the gas cell, reactor walls, and surface of packing materials. Surface analysis of packing materials was carried out by means of the scanning electron microscopy and revealed significant differences in the distribution of solid deposits on the surface of various materials. While spherical packing materials were covered by solid deposits almost uniformly, the deposits on $BaTiO_3$ and cylindrical packing materials agglomerated in characteristic circular patterns. We assume it is the result of different discharge modes (surface discharge mode vs. filamentary microdischarge mode) that form depending on the used packing material. The FTIR spectroscopy served for the identification of gaseous and solid by-products of naphthalene decomposition. In addition to main gaseous products (CO, CO_2, H_2O and $HCOOH$), several other complex gaseous and solid by-products were also identified (phthalic anhydride, maleic anhydride, 1,4- and 1,2-naphthoquinone, 1,4-benzoquinone, 1,8-naphthalic anhydride, salicylic acid, phthalide, 1-naphthalenecarboxylic acid).

Author Contributions: Conceptualization, investigation, performing the experiments, data processing, writing—original draft preparation, R.C.; performing the experiments, M.C.; SEM analysis, L.S.; supervision, investigation, writing—review and editing, K.H. All authors have read and agreed to the published version of the manuscript.

Funding: This research was funded by Slovak Research and Development Agency grant APVV-17-0382, Slovak Grant Agency VEGA 1/0419/18, Comenius University grant UK/222/2020 and foundation of Nadácia Pontis (AXA_19_09).

Conflicts of Interest: The authors declare no conflict of interest.

References

1. Mizuno, A. Generation of non-thermal plasma combined with catalysts and their application in environmental technology. *Catal. Today* **2013**, *211*, 2–8. [CrossRef]
2. Whitehead, J.C. Plasma-catalysis: The known knowns, the known unknowns and the unknown unknowns. *J. Phys. D Appl. Phys.* **2016**, *49*, 243001. [CrossRef]
3. Carreon, M.L. Plasma catalysis: A brief tutorial. *Plasma Res. Express* **2019**, *1*, 043001. [CrossRef]
4. Neyts, E.C.; Bogaerts, A. Understanding plasma catalysis through modelling and simulation—A review. *J. Phys. D Appl. Phys.* **2014**, *47*, 224010. [CrossRef]
5. Neyts, E.C.; Ostrikov, K.; Sunkara, M.K.; Bogaerts, A. Plasma Catalysis: Synergistic Effects at the Nanoscale. *Chem. Rev.* **2015**, *115*, 13408–13446. [CrossRef] [PubMed]
6. Bogaerts, A.; Zhang, Q.-Z.; Zhang, Y.-R.; Van Laer, K.; Wang, W. Burning questions of plasma catalysis: Answers by modeling. *Catal. Today* **2019**, *337*, 3–14. [CrossRef]

7. Kim, H.-H.; Ogata, A.; Futamura, S. Oxygen partial pressure-dependent behavior of various catalysts for the total oxidation of VOCs using cycled system of adsorption and oxygen plasma. *Appl. Catal. B Environ.* **2008**, *79*, 356–367. [CrossRef]
8. Mei, D.; Zhu, X.; Wu, C.; Ashford, B.; Williams, P.T.; Tu, X. Plasma-photocatalytic conversion of CO_2 at low temperatures: Understanding the synergistic effect of plasma-catalysis. *Appl. Catal. B Environ.* **2016**, *182*, 525–532. [CrossRef]
9. Tu, X.; Whitehead, J.C. Plasma-catalytic dry reforming of methane in an atmospheric dielectric barrier discharge: Understanding the synergistic effect at low temperature. *Appl. Catal. B Environ.* **2012**, *125*, 439–448. [CrossRef]
10. Parvulescu, V.I.; Magureanu, M.; Lukes, P. *Plasma Chemistry and Catalysis in Gases and Liquids*; Wiley-VCH Verlag GmbH and Co.: Weinheim, Germany, 2012; ISBN 9783527330065.
11. Pasquiers, S. Removal of pollutants by plasma catalytic processes. *Eur. Phys. J. Appl. Phys.* **2006**, *184*, 177–184. [CrossRef]
12. Hammer, T.; Kishimoto, T.; Miessner, H.; Rudolph, R. Plasma enhanced selective catalytic reduction: Kinetics of NOx-removal and byproduct formation. *SAE Trans.* **1999**, *108*, 2035–2041. [CrossRef]
13. Liu, M.; Li, J.; Liu, Z.; Zhao, Y.; Jiang, N.; Wu, Y. Improve low–temperature selective catalytic reduction of NOx with NH_3 by ozone injection. *Int. J. Plasma Environ. Sci. Technol.* **2020**, *14*, e01007. [CrossRef]
14. Van Durme, J.; Dewulf, J.; Leys, C.; Van Langenhove, H. Combining non-thermal plasma with heterogeneous catalysis in waste gas treatment: A review. *Appl. Catal. B Environ.* **2008**, *78*, 324–333. [CrossRef]
15. Vandenbroucke, A.M.; Morent, R.; De Geyter, N.; Leys, C. Non-thermal plasmas for non-catalytic and catalytic VOC abatement. *J. Hazard. Mater.* **2011**, *195*, 30–54. [CrossRef]
16. Veerapandian, S.K.P.; Leys, C.; De Geyter, N.; Moren, R. Abatement of VOCs using packed bed non-thermal plasma reactors: A review. *Catalysts* **2017**, *7*, 113. [CrossRef]
17. Whitehead, J.C. Plasma catalysis: A solution for environmental problems. *Pure Appl. Chem.* **2010**, *82*, 1329–1336. [CrossRef]
18. Liu, L.; Zhang, Z.; Das, S.; Kawi, S. Reforming of tar from biomass gasification in a hybrid catalysis-plasma system: A review. *Appl. Catal. B Environ.* **2019**, *250*, 250–272. [CrossRef]
19. Michielsen, I.; Uytdenhouwen, Y.; Pype, J.; Michielsen, B.; Mertens, J.; Reniers, F.; Meynen, V.; Bogaerts, A. CO_2 dissociation in a packed bed DBD reactor: First steps towards a better understanding of plasma catalysis. *Chem. Eng. J.* **2017**, *326*, 477–488. [CrossRef]
20. Puliyalil, H.; Lašič Jurković, D.; Dasireddy, V.D.B.C.; Likozar, B. A review of plasma-assisted catalytic conversion of gaseous carbon dioxide and methane into value-added platform chemicals and fuels. *RSC Adv.* **2018**, *8*, 27481–27508. [CrossRef]
21. Gomez-Rueda, Y.; Helsen, L. The role of plasma in syngas tar cracking. *Biomass Convers. Biorefinery* **2020**, *10*, 857–871. [CrossRef]
22. Sasujit, K.; Dussadee, N.; Tippayawong, N. Overview of tar reduction in biomass-derived producer gas using non-thermal plasma discharges. *Maejo Int. J. Sci. Technol.* **2019**, *13*, 42–61.
23. Saleem, F.; Harris, J.; Zhang, K.; Harvey, A. Non-thermal plasma as a promising route for the removal of tar from the product gas of biomass gasification—A critical review. *Chem. Eng. J.* **2020**, *382*, 122761. [CrossRef]
24. Pemen, A.J.M.; Nair, S.A.; Yan, K.; van Heesch, E.J.M.; Ptasinski, K.J.; Drinkenburg, A.A.H. Pulsed corona discharges for tar removal from biomass derived fuel gas. *Plasmas Polym.* **2003**, *8*, 209–224. [CrossRef]
25. Anis, S.; Zainal, Z.A. Tar reduction in biomass producer gas via mechanical, catalytic and thermal methods: A review. *Renew. Sustain. Energy Rev.* **2011**, *15*, 2355–2377. [CrossRef]
26. Keyte, I.J.; Albinet, A.; Harrison, R.M. On-road traffic emissions of polycyclic aromatic hydrocarbons and their oxy- and nitro-derivative compounds measured in road tunnel environments. *Sci. Total Environ.* **2016**, *566–567*, 1131–1142. [CrossRef] [PubMed]
27. Liu, L.; Wang, Q.; Ahmad, S.; Yang, X.; Ji, M.; Sun, Y. Steam reforming of toluene as model biomass tar to H_2-rich syngas in a DBD plasma-catalytic system. *J. Energy Inst.* **2018**, *91*, 927–939. [CrossRef]
28. Liu, S.; Mei, D.; Wang, L.; Tu, X. Steam reforming of toluene as biomass tar model compound in a gliding arc discharge reactor. *Chem. Eng. J.* **2017**, *307*, 793–802. [CrossRef]
29. Saleem, F.; Zhang, K.; Harvey, A. Plasma-assisted decomposition of a biomass gasification tar analogue into lower hydrocarbons in a synthetic product gas using a dielectric barrier discharge reactor. *Fuel* **2019**, *235*, 1412–1419. [CrossRef]

30. Sun, J.; Wang, Q.; Wang, W.; Song, Z.; Zhao, X.; Mao, Y.; Ma, C. Novel treatment of a biomass tar model compound via microwave-metal discharges. *Fuel* **2017**, *207*, 121–125. [CrossRef]

31. Tao, K.; Ohta, N.; Liu, G.; Yoneyama, Y.; Wang, T.; Tsubaki, N. Plasma enhanced catalytic reforming of biomass tar model compound to syngas. *Fuel* **2013**, *104*, 53–57. [CrossRef]

32. Liu, S.Y.; Mei, D.H.; Nahil, M.A.; Gadkari, S.; Gu, S.; Williams, P.T.; Tu, X. Hybrid plasma-catalytic steam reforming of toluene as a biomass tar model compound over Ni/Al$_2$O$_3$ catalysts. *Fuel Process. Technol.* **2017**, *166*, 269–275. [CrossRef]

33. Jamróz, P.; Kordylewski, W.; Wnukowski, M. Microwave plasma application in decomposition and steam reforming of model tar compounds. *Fuel Process. Technol.* **2018**, *169*, 1–14. [CrossRef]

34. Devi, L.; Nair, S.A.; Pemen, A.J.M.; Yan, K.; van Heesch, E.J.M.; Ptasinski, K.J.; Janssen, F.J.J.G. Tar removal from biomass gasification processes. In *Biomass and Bioenergy*; Nova Science Publishers: Hauppauge, NY, USA, 2006; pp. 249–274.

35. Gao, X.; Shen, X.; Wu, Z.; Luo, Z.; Ni, M.; Cen, K. The Mechanism of Naphthalene Decomposition in Corona Radical Shower System by DC Discharge. In *Electrostatic Precipitation*; Springer: Berlin/Heidelberg, Germany, 2008; pp. 713–717.

36. Nair, S.A. *Corona Plasma for Tar Removal*; Technische Universiteit Eindhoven: Eindhoven, The Netherlands, 2004; ISBN 9038626665.

37. Blanquet, E.; Nahil, M.A.; Williams, P.T. Reduction of tars from real biomass gasification by a non-thermal plasma-catalytic system. In Proceedings of the ISPC 2017, Montreal, QC, Canada, 31 July–4 August 2017.

38. Blanquet, E.; Nahil, M.A.; Williams, P.T. Enhanced hydrogen-rich gas production from waste biomass using pyrolysis with non-thermal plasma-catalysis. *Catal. Today* **2019**, *337*, 216–224. [CrossRef]

39. Hübner, M.; Brandenburg, R.; Neubauber, Y.; Röpcke, J. On the Reduction of Gas-Phase Naphthalene Using Char-Particles in a Packed-Bed Atmospheric Pressure Plasma. *Contrib. Plasma Phys.* **2015**, *55*, 747–752. [CrossRef]

40. Dors, M.; Kurzyńska, D. Tar Removal by Nanosecond Pulsed Dielectric Barrier Discharge. *Appl. Sci.* **2020**, *10*, 991. [CrossRef]

41. Liu, L.; Liu, Y.; Song, J.; Ahmad, S.; Liang, J.; Sun, Y. Plasma-enhanced steam reforming of different model tar compounds over Ni-based fusion catalysts. *J. Hazard. Mater.* **2019**, *377*, 24–33. [CrossRef]

42. Redolfi, M.; Blin-Simiand, N.; Duten, X.; Pasquiers, S.; Hassouni, K. Naphthalene oxidation by different non-thermal electrical discharges at atmospheric pressure. *Plasma Sci. Technol.* **2019**, *21*, 055503. [CrossRef]

43. Wu, Z.; Zhu, Z.; Hao, X.; Zhou, W.; Han, J.; Tang, X.; Yao, S.; Zhang, X. Enhanced oxidation of naphthalene using plasma activation of TiO$_2$/diatomite catalyst. *J. Hazard. Mater.* **2018**, *347*, 48–57. [CrossRef]

44. Kong, X.; Zhang, H.; Li, X.; Xu, R.; Mubeen, I.; Li, L.; Yan, J. Destruction of toluene, naphthalene and phenanthrene as model tar compounds in a modified rotating gliding arc discharge reactor. *Catalysts* **2019**, *9*, 19. [CrossRef]

45. Mei, D.; Liu, S.; Wang, Y.; Yang, H.; Bo, Z.; Tu, X. Enhanced reforming of mixed biomass tar model compounds using a hybrid gliding arc plasma catalytic process. *Catal. Today* **2019**, *337*, 225–233. [CrossRef]

46. Yuan, M.H.; Chang, C.C.; Chang, C.Y.; Lin, Y.Y.; Shie, J.L.; Wu, C.H.; Tseng, J.Y.; Ji, D.R. Radio-frequency-powered atmospheric pressure plasma jet for the destruction of binary mixture of naphthalene and n-butanol with Pt/Al$_2$O$_3$ catalyst. *J. Taiwan Inst. Chem. Eng.* **2014**, *45*, 468–474. [CrossRef]

47. Sutton, D.; Kelleher, B.; Ross, J.R.H. Review of literature on catalysts for biomass gasification. *Fuel Process. Technol.* **2001**, *73*, 155–173. [CrossRef]

48. Cimerman, R.; Račková, D.; Hensel, K. Tars removal by non-thermal plasma and plasma catalysis. *J. Phys. D Appl. Phys.* **2018**, *51*, 274003. [CrossRef]

49. Taheraslani, M.; Gardeniers, H. Coupling of CH$_4$ to C$_2$ hydrocarbons in a packed bed DBD plasma reactor: The effect of dielectric constant and porosity of the packing. *Energies* **2020**, *13*, 468. [CrossRef]

50. Dou, B.; Bin, F.; Wang, C.; Jia, Q.; Li, J. Discharge characteristics and abatement of volatile organic compounds using plasma reactor packed with ceramic Raschig rings. *J. Electrostat.* **2013**, *71*, 939–944. [CrossRef]

51. Mei, D.; Zhu, X.; He, Y.L.; Yan, J.D.; Tu, X. Plasma-assisted conversion of CO$_2$ in a dielectric barrier discharge reactor: Understanding the effect of packing materials. *Plasma Sources Sci. Technol.* **2015**, *24*, 15011. [CrossRef]

52. Van Laer, K.; Bogaerts, A. How bead size and dielectric constant affect the plasma behaviour in a packed bed plasma reactor: A modelling study. *Plasma Sources Sci. Technol.* **2017**, *26*, 085007. [CrossRef]

53. Wang, W.; Kim, H.H.; Van Laer, K.; Bogaerts, A. Streamer propagation in a packed bed plasma reactor for plasma catalysis applications. *Chem. Eng. J.* **2018**, *334*, 2467–2479. [CrossRef]

54. Tu, X.; Gallon, H.J.; Twigg, M.V.; Gorry, P.A.; Whitehead, J.C. Dry reforming of methane over a Ni/Al_2O_3 catalyst in a coaxial dielectric barrier discharge reactor. *J. Phys. D Appl. Phys.* **2011**, *44*, 274007. [CrossRef]

55. Tu, X.; Gallon, H.J.; Whitehead, J.C. Transition behavior of packed-bed dielectric barrier discharge in argon. *IEEE Trans. Plasma Sci.* **2011**, *39*, 2172–2173. [CrossRef]

56. Zhang, Q.Z.; Bogaerts, A. Propagation of a plasma streamer in catalyst pores. *Plasma Sources Sci. Technol.* **2018**, *27*, 035009. [CrossRef]

57. Zhang, Y.-R.; Van Laer, K.; Neyts, E.C.; Bogaerts, A. Can plasma be formed in catalyst pores? A modeling investigation. *Appl. Catal. B Environ.* **2016**, *185*, 56–67. [CrossRef]

58. Hensel, K.; Martišovitš, V.; Machala, Z.; Janda, M.; Leštinský, M.; Tardiveau, P.; Mizuno, A. Electrical and optical properties of AC microdischarges in porous ceramics. *Plasma Process. Polym.* **2007**, *4*, 682–693. [CrossRef]

59. Lee, B.Y.; Park, S.H.; Lee, S.C.; Kang, M.; Choung, S.J. Decomposition of benzene by using a discharge plasma-photocatalyst hybrid system. *Catal. Today* **2004**, *93–95*, 769–776. [CrossRef]

60. Kang, M.; Kim, B.J.; Cho, S.M.; Chung, C.H.; Kim, B.W.; Han, G.Y.; Yoon, K.J. Decomposition of toluene using an atmospheric pressure plasma/TiO_2 catalytic system. *J. Mol. Catal. A Chem.* **2002**, *180*, 125–132. [CrossRef]

61. Feng, X.; Liu, H.; He, C.; Shen, Z.; Wang, T. Synergistic effects and mechanism of a non-thermal plasma catalysis system in volatile organic compound removal: A review. *Catal. Sci. Technol.* **2018**, *8*, 936–954. [CrossRef]

62. Wallis, A.E.; Whitehead, J.C.; Zhang, K. Plasma-assisted catalysis for the destruction of CFC-12 in atmospheric pressure gas streams using TiO_2. *Catal. Lett.* **2007**, *113*, 29–33. [CrossRef]

63. Theurich, J.; Bahnemann, D.W.; Vogel, R.; Ehamed, F.E.; Alhakimi, G.; Rajab, I. Photocatalytic degradation of naphthalene and anthracene: GC-MS analysis of the degradation pathway. *Res. Chem. Intermed.* **1997**, *23*, 247–274. [CrossRef]

64. Ndifor, E.N.; Garcia, T.; Taylor, S.H. Naphthalene oxidation over vanadium-modified Pt catalysts supported on γ-Al_2O_3. *Catal. Lett.* **2006**, *110*, 125–128. [CrossRef]

65. Shie, J.L.; Chang, C.Y.; Chen, J.H.; Tsai, W.T.; Chen, Y.H.; Chiou, C.S.; Chang, C.F. Catalytic oxidation of naphthalene using a Pt/Al_2O_3 catalyst. *Appl. Catal. B Environ.* **2005**, *58*, 289–297. [CrossRef]

66. Sellick, D.R.; Morgan, D.J.; Taylor, S.H. Silica supported platinum catalysts for total oxidation of the polyaromatic hydrocarbon naphthalene: An investigation of metal loading and calcination temperature. *Catalysts* **2015**, *5*, 690–702. [CrossRef]

67. Ndifor, E.N.; Carley, A.F.; Taylor, S.H. The role of support on the performance of platinum-based catalysts for the total oxidation of polycyclic aromatic hydrocarbons. *Catal. Today* **2008**, *137*, 362–366. [CrossRef]

68. Kamal, M.S.; Razzak, S.A.; Hossain, M.M. Catalytic oxidation of volatile organic compounds (VOCs)—A review. *Atmos. Environ.* **2016**, *140*, 117–134. [CrossRef]

69. Zhang, Z.; Jiang, Z.; Shangguan, W. Low-temperature catalysis for VOCs removal in technology and application: A state-of-the-art review. *Catal. Today* **2016**, *264*, 270–278. [CrossRef]

70. Kauppi, E.I.; Honkala, K.; Krause, A.O.I.; Kanervo, J.M.; Lefferts, L. ZrO_2 Acting as a Redox Catalyst. *Top. Catal.* **2016**, *59*, 823–832. [CrossRef]

71. Yamaguchi, T. Application of ZrO_2 as a catalyst and a catalyst support. *Catal. Today* **1994**, *20*, 199–217. [CrossRef]

72. Rönkkönen, H.; Rikkinen, E.; Linnekoski, J.; Simell, P.; Reinikainen, M.; Krause, O. Effect of gasification gas components on naphthalene decomposition over ZrO_2. *Catal. Today* **2009**, *147*, 230–236. [CrossRef]

73. Viinikainen, T.; Rönkkönen, H.; Bradshaw, H.; Stephenson, H.; Airaksinen, S.; Reinikainen, M.; Simell, P.; Krause, O. Acidic and basic surface sites of zirconia-based biomass gasification gas clean-up catalysts. *Appl. Catal. A Gen.* **2009**, *362*, 169–177. [CrossRef]

74. Juutilainen, S.J.; Simell, P.A.; Krause, A.O.I. Zirconia: Selective oxidation catalyst for removal of tar and ammonia from biomass gasification gas. *Appl. Catal. B Environ.* **2006**, *62*, 86–92. [CrossRef]

75. Fang, D.; Luo, Z.; Liu, S.; Zeng, T.; Liu, L.; Xu, J.; Bai, Z.; Xu, W. Photoluminescence properties and photocatalytic activities of zirconia nanotube arrays fabricated by anodization. *Opt. Mater.* **2013**, *35*, 1461–1466. [CrossRef]

76. Fisher, J.; Egerton, T.A. Titanium Compounds, Inorganic. In *Kirk-Othmer Encyclopedia of Chemical Technology*; John Wiley and Sons: Hoboken, NJ, USA, 2001.

77. Machala, Z.; Janda, M.; Hensel, K.; Jedlovský, I.; Leštinská, L.; Foltin, V.; Martišovitš, V.; Morvová, M. Emission spectroscopy of atmospheric pressure plasmas for bio-medical and environmental applications. *J. Mol. Spectrosc.* **2007**, *243*, 194–201. [CrossRef]

78. Cheng, Z.X.; Zhao, X.G.; Li, J.L.; Zhu, Q.M. Role of support in CO_2 reforming of CH_4 over a Ni/γ-Al_2O_3 catalyst. *Appl. Catal. A Gen.* **2001**, *205*, 31–36. [CrossRef]

79. Meephoka, C.; Chaisuk, C.; Samparnpiboon, P.; Praserthdam, P. Effect of phase composition between nano γ- and χ-Al_2O_3 on Pt/Al_2O_3 catalyst in CO oxidation. *Catal. Commun.* **2008**, *9*, 546–550. [CrossRef]

80. Ogata, A.; Yamanouchi, K.; Mizuno, K.; Kushiyama, S.; Yamamoto, T. Decomposition of benzene using alumina-hybrid and catalyst-hybrid plasma reactors. *IEEE Trans. Ind. Appl.* **1999**, *35*, 1289–1295. [CrossRef]

81. Sano, T.; Negishi, N.; Sakai, E.; Matsuzawa, S. Contributions of photocatalytic/catalytic activities of TiO_2 and γ-Al_2O_3 in nonthermal plasma on oxidation of acetaldehyde and CO. *J. Mol. Catal. A Chem.* **2006**, *245*, 235–241. [CrossRef]

82. Roland, U.; Holzer, F.; Kopinke, F.D. Combination of non-thermal plasma and heterogeneous catalysis for oxidation of volatile organic compounds: Part 2. Ozone decomposition and deactivation of γ-Al_2O_3. *Appl. Catal. B Environ.* **2005**, *58*, 217–226. [CrossRef]

83. Ray, D.; Manoj Kumar Reddy, P.; Subrahmanyam, C. Glass Beads Packed DBD-Plasma Assisted Dry Reforming of Methane. *Top. Catal.* **2017**, *60*, 869–878. [CrossRef]

84. Holzer, F.; Kopinke, F.D.; Roland, U. Influence of ferroelectric materials and catalysts on the performance of Non-Thermal Plasma (NTP) for the removal of air pollutants. *Plasma Chem. Plasma Process.* **2005**, *25*, 595–611. [CrossRef]

85. Kim, H.-H.; Teramoto, Y.; Negishi, N.; Ogata, A. A multidisciplinary approach to understand the interactions of nonthermal plasma and catalyst: A review. *Catal. Today* **2015**, *256*, 13–22. [CrossRef]

86. Butterworth, T.; Elder, R.; Allen, R. Effects of particle size on CO_2 reduction and discharge characteristics in a packed bed plasma reactor. *Chem. Eng. J.* **2016**, *293*, 55–67. [CrossRef]

87. Xu, S.; Whitehead, J.C.; Martin, P.A. CO_2 conversion in a non-thermal, barium titanate packed bed plasma reactor: The effect of dilution by Ar and N_2. *Chem. Eng. J.* **2017**, *327*, 764–773. [CrossRef]

88. Liu, Y.; Li, Z.; Yang, X.; Xing, Y.; Tsai, C.; Yang, Q.; Wang, Z.; Yang, R.T. Performance of mesoporous silicas (MCM-41 and SBA-15) and carbon (CMK-3) in the removal of gas-phase naphthalene: Adsorption capacity, rate and regenerability. *RSC Adv.* **2016**, *6*, 21193–21203. [CrossRef]

89. Ravenni, G.; Sárossy, Z.; Ahrenfeldt, J.; Henriksen, U.B. Activity of chars and activated carbons for removal and decomposition of tar model compounds—A review. *Renew. Sustain. Energy Rev.* **2018**, *94*, 1044–1056. [CrossRef]

90. Bogaerts, A.; Kozak, T.; Van Laer, K.; Snoeckx, R. Plasma-based conversion of CO_2: Current status and future challenges. *Faraday Discuss.* **2015**, *183*, 217–232. [CrossRef] [PubMed]

91. Van Laer, K.; Bogaerts, A. Improving the Conversion and Energy Efficiency of Carbon Dioxide Splitting in a Zirconia-Packed Dielectric Barrier Discharge Reactor. *Energy Technol.* **2015**, *3*, 1038–1044. [CrossRef]

92. Whitehead, J.C. Plasma-catalysis: Is it just a question of scale? *Front. Chem. Sci. Eng.* **2019**, *13*, 264–273. [CrossRef]

93. Wang, Y.; Yang, H.; Tu, X. Plasma reforming of naphthalene as a tar model compound of biomass gasification. *Energy Convers. Manag.* **2019**, *187*, 593–604. [CrossRef]

94. Zhang, Q.-Z.; Wang, W.-Z.; Bogaerts, A. Importance of surface charging during plasma streamer propagation in catalyst pores. *Plasma Sources Sci. Technol.* **2018**, *27*, 065009. [CrossRef]

95. Takaki, K.; Urashima, K.; Chang, J.S. Ferro-electric pellet shape effect on C_2F_6 removal by a packed-bed-type nonthermal plasma reactor. *IEEE Trans. Plasma Sci.* **2004**, *32*, 2175–2183. [CrossRef]

96. Smith, B. *Infrared Spectral Interpretation: A Systematic Approach*; CRC Press: Boca Raton, FL, USA, 1999; ISBN 0849324637.

97. Socrates, G. *Infrared and Raman Characteristic Group Frequencies Contents*; John Wiley and Sons: Hoboken, NJ, USA, 2001; ISBN 0471852988.

98. Bellamy, L.J. *The Infrared Spectra of Complex Molecules*; John Wiley and Sons: Hoboken, NJ, USA, 1976; ISBN 9789401160173.

99. Silverstein, R.M.; Webster, F.X.; Kiemle, D.J. *Spectrometric Identification of Organic Compounds*; John Wiley and Sons: Hoboken, NJ, USA, 2005; ISBN 0-471-39362-2.

100. Xu, R.; Zhu, F.; Zhang, H.; Ruya, P.M.; Kong, X.; Li, L.; Li, X. Simultaneous Removal of Toluene, Naphthalene, and Phenol as Tar Surrogates in a Rotating Gliding Arc Discharge Reactor. *Energy Fuels* **2020**, *34*, 2045–2054. [CrossRef]

101. Zhang, H.; Zhu, F.; Li, X.; Xu, R.; Li, L.; Yan, J.; Tu, X. Steam reforming of toluene and naphthalene as tar surrogate in a gliding arc discharge reactor. *J. Hazard. Mater.* **2019**, *369*, 244–253. [CrossRef] [PubMed]

102. Ni, M.J.; Shen, X.; Gao, X.; Wu, Z.L.; Lu, H.; Li, Z.S.; Luo, Z.Y.; Cen, K.F. Naphthalene decomposition in a DC corona radical shower discharge. *J. Zhejiang Univ. Sci. A* **2011**, *12*, 71–77. [CrossRef]

103. Wu, Z.; Wang, J.; Han, J.; Yao, S.; Xu, S.; Martin, P. Naphthalene Decomposition by Dielectric Barrier Discharges at Atmospheric Pressure. *IEEE Trans. Plasma Sci.* **2016**, *45*, 154–161. [CrossRef]

104. Abdelaziz, A.A.; Seto, T.; Abdel-Salam, M.; Otani, Y. Influence of N_2/O_2 mixtures on decomposition of naphthalene in surface dielectric barrier discharge based reactor. *Plasma Chem. Plasma Process.* **2014**, *34*, 1371–1385. [CrossRef]

105. Wypych, A.; Bobowska, I.; Tracz, M.; Opasinska, A.; Kadlubowski, S.; Krzywania-Kaliszewska, A.; Grobelny, J.; Wojciechowski, P. Dielectric properties and characterisation of titanium dioxide obtained by different chemistry methods. *J. Nanomater.* **2014**, *2014*, 124814. [CrossRef]

106. Butterworth, T.; Allen, R.W.K. Plasma-catalyst interaction studied in a single pellet DBD reactor: Dielectric constant effect on plasma dynamics. *Plasma Sources Sci. Technol.* **2017**, *26*, 065008. [CrossRef]

107. Gomaa, M.M.; Gobara, H.M. Electrical properties of Ni/silica gel and Pt/γ-alumina catalysts in relation to metal content in the frequency domain. *Mater. Chem. Phys.* **2009**, *113*, 790–796. [CrossRef]

108. Harrop, P.J.; Wanklyn, J.N. The dielectric constant of zirconia. *Br. J. Appl. Phys.* **1967**, *18*, 739–742. [CrossRef]

Publisher's Note: MDPI stays neutral with regard to jurisdictional claims in published maps and institutional affiliations.

 *catalysts*

Article

Plasma-Catalysis for Volatile Organic Compounds Decomposition: Complexity of the Reaction Pathways during Acetaldehyde Removal

Arlette Vega-González *[ID], Xavier Duten and Sonia Sauce

CNRS, UPR 3407, Laboratoire des Sciences des Procédés et des Matériaux, LSPM, Université Sorbonne Paris Nord, F-93430 Villetaneuse, France; xavier.duten@lspm.cnrs.fr (X.D.); sauce.sonia@gmail.com (S.S.)
* Correspondence: arlette.vega@lspm.cnrs.fr; Tel.: +33-1-49-40-34-31

Received: 4 September 2020; Accepted: 30 September 2020; Published: 3 October 2020

Abstract: Acetaldehyde removal was carried out using non-thermal plasma (NTP) at 150 J·L^{-1}, and plasma-driven catalysis (PDC) using Ag/TiO$_2$/SiO$_2$, at three different input energies—70, 350 and 1150 J·L^{-1}. For the experimental configuration used, the PDC process showed better results in acetaldehyde (CH$_3$CHO) degradation. At the exit of the reactor, for both processes and for all the used energies, the same intermediates in CH$_3$CHO decomposition were identified, except for acetone which was only produced in the PDC process. In order to contribute to a better understanding of the synergistic effect between the plasma and the catalyst, acetaldehyde/catalyst surface interactions were studied by diffuse-reflectance infrared Fourier transform spectroscopy (DRIFTS). These measurements showed that different species such as acetate, formate, methoxy, ethoxy and formaldehyde are present on the surface, once it has been in contact with the plasma. A reaction pathway for CH$_3$CHO degradation is proposed taking into account all the identified compounds in both the gas phase and the catalyst surface. It is very likely that in CH$_3$CHO degradation the presence of methanol, one of the intermediates, combined with oxygen activation by silver atoms on the surface, are key elements in the performance of the PDC process.

Keywords: non-thermal plasma; catalysis; VOC; DRIFTS; acetaldehyde

---

1. Introduction

Diphasic processes combining a non-thermal plasma at atmospheric pressure and a catalytic bed have become of major interest for air-pollutant removal, and particularly volatile organic compounds (VOC) [1–3]. Indeed, compared to classical air cleaning techniques (thermal catalysis or photocatalysis), plasma-driven catalysis can lead to similar or better VOC degradation rates with less energy injected into the reactor and without catalyst deactivation issues [4,5]. Moreover, in the case of indoor air cleaning, where low VOC concentrations (in the ppm order) have to be removed, this diphasic process seems to be the most suitable [6]. Reviews dealing with the synergistic effects and mechanism of a non-thermal plasma catalysis system in volatile organic compound removal have recently been published [7–9]. Among the studies dealing with the different aspects of such a complex process, we can mention studies on plasma/pollutant interactions [10], surface charging [11,12], adsorption [13,14] or catalyst activation by non-thermal plasma (NTP) [15].

Given the fact that the induced heterogeneous reactions and the plasma are interdependent due to the plasma's continuous discharge during processing, it is essential to make further progress on identifying the surface species forming on the catalyst under the plasma effect. The in-situ Fourier transformed infrared spectroscopy (FTIR)/diffuse-reflectance infrared Fourier transform spectroscopy (DRIFTS) approach for studying catalysts under working conditions is gaining importance in the framework of the plasma-driven catalysis (PDC) process. To our knowledge, the first of these studies was carried out on isopropanol conversion on γ-Al_2O_3 [16]. The authors used a dielectric barrier discharge (DBD) plasma reactor, under static conditions, coupled to in-situ FTIR spectroscopy. They were able to show that unlike the thermal catalysis pathway, with the DBD process the first reaction intermediate is acetone. Plasma-assisted hydrocarbon selective catalytic reduction [17], catalytic steam-reforming of methane [18], reverse water–gas [19]/water–gas shift [20] and plasma-assisted CO_2 hydrogenation [21] processes have also been studied using in-situ DRIFTS. These works pointed out the fact that NTP is able to activate some of the species present either in the gas phase or on the surface, thus improving the performance of the PDC process. Indeed, the role of oxygen in the PDC process has been investigated via isotopically-labeled molecular oxygen ($^{18}O_2$); the results showed that oxygen was fixed onto the surface of the catalyst by the action of the NTP discharge, and was able to survive in such a state for about 30 min [22]. What is more, this same study demonstrated that when using Ag nanoparticles supported on catalysts, they served as oxygen reservoirs. As for VOC removal with non-thermal plasma-assisted catalysis, isopropanol [23,24], toluene [24,25] and acetone removal [26] have been studied using an FTIR in-situ technique. This approach allowed the identification of reaction intermediates, and the following of the evolution of secondary compounds arising from the VOC oxidation.

In a previous work [27] we have showed that, by combining a DBD generated at atmospheric pressure and a silver-supported nanostructured catalyst, acetaldehyde can be removed up to 98% with the production of mainly CO, CO_2 and O_3. In addition, we have also studied acetaldehyde adsorption and ozonation over the same silver-based catalyst, and the obtained results support the fact that PDC is a rather complex process in which homogeneous and heterogeneous chemistry are closely interconnected [28]. In the present study we analyse the changes that occurred on an $Ag/TiO_2/SiO_2$ catalyst, used for acetaldehyde removal, after plasma exposure. We investigate which intermediate species were formed on the surface after acetaldehyde adsorption and discharge-assisted conversion by using DRIFT spectroscopy. Gaseous by-products resulting from this process were monitored and quantified, with the aim to contribute to a better understanding of the mechanism of acetaldehyde oxidation. Essentially based on the reactor outlet analysis, the surface composition and literature data, an acetaldehyde removal pathway is proposed considering both the homogeneous and heterogeneous chemistry of the process.

2. Results

The results obtained when studying acetaldehyde removal using non-thermal plasma, NTP-plasma driven catalysis or just a catalyst will be briefly presented. In addition, we have previously shown that VOC adsorption and heterogeneous NTP-induced chemistry have to be taken into account when associating NTP in air with a catalyst. These results, discussed in more detail in previous works [27–29], were obtained at ambient temperature and atmospheric pressure, but in different configurations of the plasma discharge (i.e., corona or DBD discharge, diameter of the catalyst support, mass of catalyst, energy injected).

2.1. Acetaldehyde Decomposition Using the Plasma-Driven Catalysis Process

The results obtained for acetaldehyde decomposition with the single stage plasma-driven catalysis, and the plasma-alone processes are presented in Table 1. It is worth noting that the degradation of acetaldehyde by the catalyst alone in the DBD reactor was also studied at room temperature and atmospheric pressure. In this case, neither any degradation of acetaldehyde nor by-product formation is observed.

As can be seen from the results presented in the table above, at atmospheric pressure and ambient temperature the catalyst is not active for acetaldehyde removal. With the plasma-alone process acetaldehyde can be removed up to 55% with an energy consumption of 150 J·L^{-1} (~250 mW power consumption). As soon as the silver nanostructured catalyst is introduced into the discharge zone, acetaldehyde removal increases, already allowing, at 70 J·L^{-1} (~120 mW), 33% to be obtained and up to 98% at 1150 J·L^{-1} (~2 mW).

Table 1. Acetaldehyde removal obtained using the plasma-driven catalysis and the non-thermal plasma (NTP) processes.

Specific Input Energy (SIE) (J·L^{-1})	Ag/TiO$_2$/SiO$_2$	Plasma Alone	Plasma + Ag/TiO$_2$/SiO$_2$		
	Alone	150	70	350	1150
Acetaldehyde removal (%)	0	55	33	87	98

The main gaseous by-products at the exit of the DBD reactor are CO and CO$_2$, but other organic by-products have also been identified: methanol (Me), acetic acid (AcA), acetone (Ace), methyl formate (MeF), methyl acetate (MeA), 1,2-ethanediol monoformate (EmF), 1,2-ethanediol diformate (EdF), nitromethane (Nm), methyl nitrate (MN) and 2-ethoxyethanol (Eet). Their distribution at the DBD reactor exit is presented in Figure 1, except for the ones present in trace amounts (Eet), or those that could not be quantified (MN). Among these organic products we recognize compounds belonging to different functional groups such as hydroxyl, carboxyl, carbonyl, carboalkoxy and nitro compounds, reflecting the complexity of the chemistry involved in such processes.

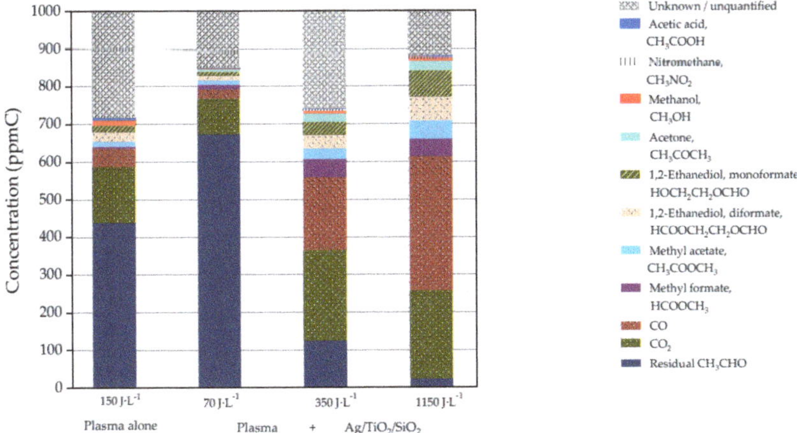

Figure 1. Comparison of the composition of the dielectric barrier discharge (DBD) reactor effluent obtained with the plasma-alone and the plasma-driven catalysis process, as a function of SIE.

From the obtained results we can say that with the plasma-alone process neither acetone nor nitromethane are produced at 150 J·L^{-1}; and compared with the PDC process, more acetic acid and methanol are formed. With the PDC process at 70 J·L, aside from methyl formate and the emergence of acetone, all the other by-products are produced in lower amounts compared with NTP alone, which is certainly due to the lower acetaldehyde conversion. The by-products' formation is promoted as soon as the specific input energy (SIE) increases. Between 70 and 350 J·L^{-1}, there is a significant rise in each by-product concentration, 150% being the lowest increase observed. Between 350 and 1150 J·L^{-1}, an almost constant production of CO$_2$, methanol, acetone, methyl formate and nitromethane is observed; whereas an increase between 70/80% in CO, MeA, AcA and EdF, and of 108% in EmF is still observed. Even if the amount of MN could not be quantified, the analyses of the DBD reactor effluent have shown that it is formed in higher quantities in the PDC process, and its production increases with SIE. Furthermore, introducing a catalyst in the discharge zone induces an increase in CO$_x$ selectivity. In fact, NTP at 150 J·L^{-1} has a CO$_x$ selectivity of 35% while in the PDC process this selectivity is equal to 36, 50 and 60% for 70, 350 and 1150 J·L^{-1}, respectively.

2.2. Acetaldehyde Adsorption on Ag/TiO$_2$/SiO$_2$

In order to characterize the catalyst surface before turning the plasma on, in-situ DRIFTS was used to monitor the Ag/TiO$_2$/SiO$_2$ surface during acetaldehyde adsorption. The adsorption capacity of the catalyst, at the experimental conditions used, was evaluated to be 698 ± 3 μmol/g$_{catalyst}$.

Figure 2 shows the infrared spectra of the silver-based catalyst surface after 1 and 30 min of exposition to 1000 ppmC of CH$_3$CHO at 298 K and atmospheric pressure. Acetaldehyde shows peaks in the 3000–2600 cm^{-1} region and 1800–1500 cm^{-1} region, as shown in the figure. After 1 min acetaldehyde adsorption, bands at 2936, 2922, 2866, 2770, 2748, 1759, 1713, 1555 and 1242 cm^{-1} are observed. Once the saturation is attained, after 30 min, the previously mentioned absorption bands increase, and a new band appears at 2974 cm^{-1}. In addition, broadband absorption features, associated with the peaks at 1713, 1555 and 1242 cm^{-1} are observed. The first broadband presents peaks at 1720 and 1713 cm^{-1}; a small shoulder at 1697 cm^{-1}, and a large shoulder in the 1690–1650 cm^{-1} range. The second broadband still showing a maximum at 1555 cm^{-1}, and many smaller peaks in the 1590–1500 cm^{-1} range are visible after saturation of the surface with acetaldehyde. The last broadband also presents smaller peaks in the 1280–1210 cm^{-1} range.

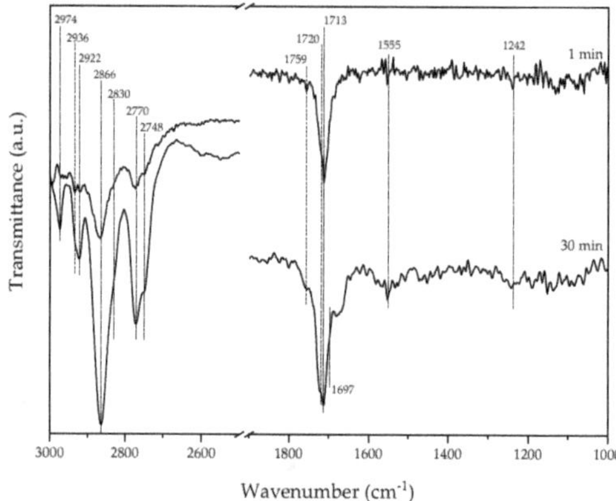

Figure 2. Diffuse-reflectance infrared Fourier transform spectroscopy (DRIFTS) spectra recorded after 1 and 30 min of acetaldehyde adsorption on Ag/TiO$_2$/SiO$_2$ at 298 K and atmospheric pressure.

Peak assignments are presented in Table 2. In both spectra, bands in the 3000–2800 cm^{-1} region, the CH stretching region, can be assigned to symmetric and asymmetric ν(CH$_3$) and ν(CH$_2$) modes [30] and bands in the 1700–1680 cm^{-1} region can be assigned to ν(C=O) mode [31].

Table 2. Adsorbed acetaldehyde: observed bands and their assignments.

Vibrational Mode	Vibrational Assignments (cm^{-1})		Reference
	Gas Phase CH$_3$CHO	CH$_3$CHO Adsorbed on Ag/TiO$_2$/SiO$_2$	
ν_{as}(CH$_3$)	2967	2974	[30]
ν_{as}(CH$_2$)		2936	[30–32]
ν_s(CH$_3$)	2923	2922	[30,33]
2ν_6A′ Fermi	2840	2866	[34]
ν_s(CH$_2$)		2830	[30]
ν(CH)η^1-acetaldehyde	2736, 2704	2770, 2748	[31,33,35]
ν(C=O)	1735	1713	[31]
ν_{as}(COO)		1555	[36]

From the literature, most of the bands in the CH stretching region could be assigned to adsorbed acetaldehyde [32–34]. However significant overlap of much of this region exists between acetaldehyde and the products of its heterogeneous reactions on oxide surfaces. Indeed, the bands at 2922, 2830 and 2748 cm^{-1} could be assigned to crotonaldehyde [33] whereas the band at 2936 cm^{-1} may also be assigned to other condensation products derived from surface reactions [31], or to acetate species [32]. The formation of crotonaldehyde is supported by the presence of the peak at 1720 cm^{-1} and the broad shoulder at 1690–1650 cm^{-1} [31,33,37–39] and the formation of acetate species is in agreement with the emergence of the broadband centered at 1555 cm^{-1} [36]. The dominant spectral feature in the $\nu(C=O)$ region, at 1713 cm^{-1}, which is assigned to adsorbed acetaldehyde, shifts to lower wavenumbers with respect to the same band of CH_3CHO in the gas phase. This indicates that acetaldehyde is bound to the surface TiO_2 through its carbonyl group by H-bridge bonding [37,38,40]. In addition, the growth of other infrared spectral features in this region points out the probable formation of other products such as acetone (1697 cm^{-1} [41]) or formaldehyde. Indeed, the band at 1242 cm^{-1} can be assigned to molecularly adsorbed formaldehyde [42,43], which has also been reported to present bands at 2913, 2863, 2759, 1648 and 1413 cm^{-1} [42,44], supporting the presence of formaldehyde on the silver-based catalyst surface after acetaldehyde adsorption.

Although at the end of the adsorption step the $Ag/TiO_2/SiO_2$ catalyst surface is mainly covered with adsorbed acetaldehyde, the other species which are also present have to be considered in order to better understand the acetaldehyde degradation mechanism in the following step of the process.

2.3. Surface Species Formed during the PDC Process

Once saturated with acetaldehyde, the $Ag/TiO_2/SiO_2$ catalyst is used in order to decompose the VOC with the PDC process. Three different energies were used—70, 350 and 1150 $J{\cdot}L^{-1}$. Figure 3 shows the surface spectra of the catalyst before and after being exposed to the DBD discharge for 30 min. It is clear that the surface species present on the surface after the adsorption process take part in different surface reactions while the plasma is on, leading to the modification of the surface composition as can be seen from the FTIR band features.

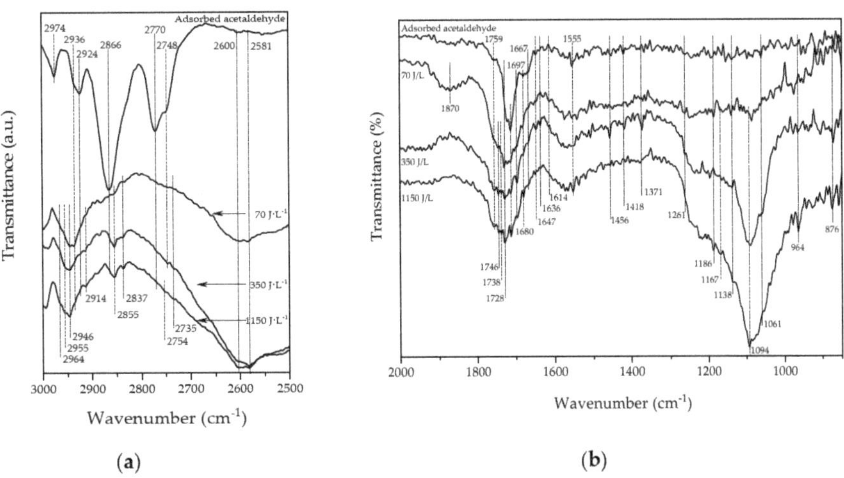

Figure 3. $Ag/TiO_2/SiO_2$ catalyst surface DRIFTS spectra after saturation with CH_3CHO; and NTP effect on surface species at different plasma discharge energies: (**a**) C-H stretching region; (**b**) bending region.

From Figure 3a it is clear that, after the PDC process, some of the bands disappear (2974 and 2770 cm^{-1}) or almost disappear (2748 cm^{-1}). The relative intensity of the bands in the 2970–2900 cm^{-1} region has clearly changed, with visible peaks at 2946, 2955 and 2964 cm^{-1}; the band in the 2900–2800 cm^{-1} region decreases significantly; and a new broad band with peaks at 2600 and 2581 cm^{-1} appears. Figure 3b shows an evident broadening of the bands centered at 1713, 1555 and

1242 cm^{-1} in the acetaldehyde-saturated surface. As the plasma input energy increases, the maximum of the v(C=O) band is redshifted and its intensity decreases for an SIE higher than 70 J·L^{-1}. At the same time, a band with increasing intensity develops in the 1300–1000 cm^{-1} region.

The loss of the bands at 2974 and 2770 cm^{-1} (assigned to adsorbed acetaldehyde), along with an intensity decrease of the peaks at 2866 and 1713 cm^{-1} indicates a gradual depletion of acetaldehyde surface species once the catalyst is in contact with the plasma. Likewise, the loss in intensity of the bands at 2922, 2830, 2748 and 1720 cm^{-1} indicates that less crotonaldehyde is present on the surface. Under the plasma effect, the 2950–2900 cm^{-1} band broadens, and presents different small peaks and shoulders with maximum intensities at 2946 and 2936 cm^{-1}. The former peak may be assigned to surface methoxy [45]; indeed, along with the peak at 2946 cm^{-1}, the presence of a small shoulder at 2924 cm^{-1} and a peak at 2855 cm^{-1} supports the possibility of the formation of surface methoxy [45–47]. Concerning the peak at 2936 cm^{-1}, its attribution is less straightforward as it can be related to different surface species such as formate, ethoxy or acetate species [32,44,45]. On the one hand, the infrared features at 2964, 2955, 2936, 2924, 2866 and 2735 cm^{-1} bring out the formation of surface formate [45,48]. On the other hand, the peaks at 2964, 2936, 2866 and 2855 cm^{-1} have also been assigned to ethoxy species formed on a TiO$_2$ surface [37,44,49]. Furthermore, the peaks at 2964 and 2936 cm^{-1} have also been assigned to methyl acetate [35], whereas the one at 2936 cm^{-1} is also assigned to acetic acid [50]. The broad band between 2650 and 2550 cm^{-1} that appears after the plasma catalysis process supports the formation of formate species as it has been attributed to formic acid [51]. As the energy increases up to 350 J/L, the 2900–2800 cm^{-1} band loses intensity, the loss being more pronounced for the peak at 2936 cm^{-1}; whereas the peaks at 2964 and 2955 cm^{-1}, as well as a new peak at 2914 cm^{-1} become more visible. This latter peak may be attributed either to acetaldehyde [36], or to an intermediate formed from formaldehyde oxidation or disproportionation to give formate or methoxide, respectively [52,53].

These attributions can be backed by the infrared features in the 2000–950 cm^{-1} region. Significant overlapping of the peaks in this wavelength range complicates precise identification of surface species but, based on litterature data, the bands generated on the plasma-exposed surface can be related to distinct vibrational modes of acetate, formate, methoxy, ethoxy, carbonate and formaldehyde species adsorbed on the surface of the Ag/TiO$_2$/SiO$_2$ catalyst. Thus, we may assign some of the bands as presented in Table 3.

Table 3. Infrared (IR) bands assignment of the major surface species formed during the PDC process, in the 2000–950 cm^{-1} region.

Surface Species	IR Band (cm^{-1})	Reference
Acetate	1560–1540/1450–1418	[33,39,46,54–56]
Formate	1870–1828/1590–1550/1380–1350	[42,44,45,47,54,55,57–60]
Methoxy	964/1100–1030	[61,62]
Ethoxy	1456/1380–1390/1190–1090/1065–1050	[37,49,56,63]
Carbonate	1680/1614/1520/1430/1310	[42]
Acetic acid	1736/1675/1535/1453/1415/1341/1296/1025–1050	[44,46,56]
Formaldehyde	1767–1746/1727–1713/1418/1260	[42,43,52,55]

The change in the vibrational spectrum of the catalyst surface following its exposure to NTP suggests a modification of its composition, which is directly related to the acetaldehyde decomposition mechanism.

3. Discussion

In this section we will propose a reaction pathway that takes into account acetaldehyde adsorption and its degradation on the surface with the PDC process. On the basis of the DBD reactor effluent analysis, and the infrared spectroscopic measurements carried out on the catalyst surface, it is evident that acetaldehyde undergoes a rather complex chemistry during the process. For this reason, a detailed description of the acetaldehyde decomposition mechanism is out-of-reach at this point. Alternatively, we propose a series of simplified adsorption/degradation routes in order to account for the key aspects of surface activity that lead to the formation of the different identified surface species, and a subsequent increase in acetaldehyde degradation, which results in an increase in the amount of gaseous by-products.

3.1. Acetaldehyde Adsorption Mechanism

In different studies concerning the photocatalytic degradation pathway of acetaldehyde on a TiO_2 surface, it has been noticed that, upon adsorption of acetaldehyde, different species are already formed without illuminating the catalyst. Two compounds, 3-hydroxybutanal and crotonaldehyde, are formed through an aldol condensation, while bidentate acetate species have been detected to be formed by an oxidation process occurring on the TiO_2 surface [40].

As suggested from the spectra analysis of the saturated surface, when acetaldehyde is brought into contact with the nanostructured TiO_2, it will certainly bind to the surface through the oxygen atom of its carbonyl group. Then, two acetaldehyde molecules adsorbed on neighboring sites can react through a β-aldolization reaction to give crotonaldehyde, $CH_3(CH)_2CHO$ [28,31,33,64–66]. It has been shown that acetaldehyde undergoes aldol condensation forming 3-hydroxybutanal ($CH_3CH(OH)CH_2CHO$) which, upon dehydration forms crotonaldehyde [31,34,40]:

$$CH_3CHO_{ads} + CH_3CHO_{ads} \rightarrow CH_3 - CH = CH - CHO_{ads} + H_2O \tag{1}$$

Formation of acetate species can result from the direct oxidation on the surface of acetaldehyde [30, 32,46,66], and to a lesser extent of crotonaldehyde [40], through a surface oxygen (O_s):

$$CH_3CHO_{ads} + O_s \rightarrow CH_3COO^-_{ads} + H^+_{ads} \tag{2}$$

Besides, a Cannizzaro disproportionation of two acetaldehyde molecules involving a vacancy (V_O) and surface oxygen would result in the formation of acetate and ethoxy groups. This pathway has been proposed for acetaldehyde adsorption on rutile TiO_2 [33]:

$$2CH_3CHO_{ads} + V_O/O_s \rightarrow CH_3COO^-_{ads} + CH_3CH_2O^-_{ads} \tag{3}$$

Despite the fact that it is difficult to ascertain the presence of ethoxy species from the spectra in Figure 2, their presence at much smaller levels than acetates cannot be completely ruled out.

Concerning acetone formation on oxide surfaces from aldehydes, two pathways are mainly known. The first is via a two-step reaction: oxidation to carboxylates followed by the coupling of two carboxylates to form one molecule of ketone. The second pathway is through the reaction of an adsorbed acyl group, RC=O, with an alkyl group [32]. Examples of acetone formation from acetaldehyde via these two pathways have been reported, but they concern studies carried out at temperatures well above the ambient temperature [32,67–69]. Another possibility is acetone formation through 3-hydroxybutanal. Indeed, at ambient temperature this compound is very reactive and, either it undergoes fast dehydration on the surface to form crotonaldehyde; or it can be converted via intramolecular H transfer to 4-hydroxybutan-2-al ($CH_3C(O)CH_2CH_2OH$). The latter can be further transferred into acetone ($CH_3C(O)CH_3$) and formaldehyde ($HC(O)H$) via reverse reaction to aldol condensation [34]:

$$2CH_3CHO_{ads} \rightarrow CH_3CH(OH)CH_2CHO_{ads} \leftrightarrow CH_3C(O)CH_2CH_2OH \rightarrow HC(O)H + CH_3C(O)CH_3 \tag{4}$$

This mechanism is a very plausible one, as it explains acetone and formaldehyde formation after acetaldehyde adsorption at ambient temperature.

3.2. Surface Species Formed during the PDC Process: Formation Mechanism

Before turning the plasma on, the catalyst surface is mainly composed of adsorbed acetaldehyde. Part of this acetaldehyde yields crotonaldehyde (Reaction (1)); acetate (Reactions (2) and (3)), acetone and formaldehyde (Reaction (4)). DRIFTS measurements carried out on the catalyst exposed to the plasma reveal that acetaldehyde degradation proceeds with the formation of different surface compounds such as acetate, methoxy, ethoxy, formate and carbonate species, as well as formaldehyde and formic and acetic acid. It should be noted that, despite being referred to as formic acid and acetic acid, it is most likely formate and acetate species on the surface [40]. Indeed, previous studies have shown that photocatalytic oxidation of acetaldehyde could yield acetic and formic acid that remain adsorbed on the surface in a dissociated form, i.e., as acetate and formate, respectively (degradation routes Reactions (5) and (6)) [55,69].

$$CH_3CHO_a \rightarrow CH_3COO_a^- + H_a^+ \tag{5}$$

$$CH_3CHO_a \rightarrow HCHO_a \rightarrow HCOO_a^- + H_a^+ \tag{6}$$

The formation of acetate species on the catalyst surface has already been presented in the previous section. Nevertheless, once the plasma is turned on, two more routes may be considered. One route involves the acetyl radical resulting from adsorbed acetaldehyde decomposition by the plasma; which in turn can react with gas-phase oxygen to form acetate species (Reaction (7)) [56]. The other one is the reaction of gaseous acetaldehyde with an activated surface oxygen (Reaction (8)) [70]—indeed, it has been shown that the presence of silver atoms enhances the adsorption of oxygen on the catalyst and promotes its activation [22,71].

$$CH_3CHO_a \rightarrow CH_3CO_a + H_a \xrightarrow{+O_2} CH_3COO_a^- \tag{7}$$

$$CH_3OH_g + O_a \rightarrow CH_3COO_a^- + H_a^+ \tag{8}$$

Regarding formate species, its formation through oxidation of different compounds such as acetone (Reaction (9)) [55], acetate species (Reactions (10) and (11)) [46,55], formaldehyde (Reaction (12)) [55,72–74] and crotonaldehyde (Reaction (13)) [64] should be considered.

$$CH_3C(O)CH_3\,_a \xrightarrow{O_2} 2HCOO_a^- + H_a^+ + CO_2 \tag{9}$$

$$CH_3COO_a^- + H^+ \rightarrow HCOO_a^- + H_a^+ \tag{10}$$

$$CH_3COO_a^- \rightarrow CH_3O_a \rightarrow CH_2O_a \rightarrow HCOO_a^- \tag{11}$$

$$HCHO + O_a \rightarrow HCOO_a^- \tag{12}$$

$$CH_3(CH)_2CHO_a \rightarrow HCOO_a^- + H_a^+ \tag{13}$$

Methoxy species may be formed following different routes, one of which involves methanol, a by-product of acetaldehyde decomposition by NTP. The proposed routes are direct oxidation of surface acetate species (Reaction (14)) [46], and dissociative adsorption of methanol (Reaction (15)) [75,76]. Indeed, on the basis of reaction mechanisms derived from model studies carried out under ultrahigh vacuum conditions on noble metal surfaces, methanol is activated by surface oxygen so that adsorbed methoxy species can form [77].

$$CH_3COO_a^- \rightarrow CH_3O_a + CO_{2\,(g)} \tag{14}$$

$$CH_3OH_g + S \rightarrow CH_3O_a + H_a^+ \tag{15}$$

where S is a site on the oxide surface.

Concerning ethoxy species, one possible pathway has already been proposed through the Reaction (3) route. Moreover, it has been shown that TiO_2-supported catalysts may significantly promote the reaction of acetaldehyde with adsorbed hydrogen formed in previous steps to form ethoxy species [54,66] (Reaction (16)).

$$CH_3CHO_a + H_a^+ \rightarrow CH_3CH_2O_a^- \tag{16}$$

As for carbonate species, they may be produced from acetates (Reactions (17) and (18)) [56,67,70]. Furthermore, an investigation carried out using supported TiO_2 catalysts in the presence of H_2O_2 [58] has shown that formate species interaction with the surface can lead to carbonate and acetone formation (Reaction (19)). This latter possibility should be considered in our case, as NTP is able to generate H_2O_2.

$$CH_3COO_a^- + OH_s \rightarrow CO_{3\,a}^{2-} + CH_4 \tag{17}$$

$$2CH_3COO_a^- \rightarrow CH_3C(O)CH_{3\,a} + CO_{3\,a}^{2-} \tag{18}$$

$$HCOO_a^- \xrightarrow{H_2O_2} CO_{3\,a}^{2-} \tag{19}$$

Regarding formaldehyde formation, in addition to Reaction (4) route, other routes may also be allowed for. These include formation of formaldehyde from acetaldehyde on TiO_2 surfaces, either directly or through an acetate intermediary (Reaction (20)) [55], from acetaldehyde oxidation by surface oxygen (Reaction (21)) [54], from surface ethoxy and methoxy species oxidation (Reactions (22) and (23)) [44,46,57] and as an intermediate in methanol oxidation (Reaction (24)) [74].

$$CH_3CHO_a \xrightarrow{(CH_3COO_a^-)} HCHO_a + HCOOH_a \tag{20}$$

$$CH_3CHO_a + 2O_s \rightarrow HCHO_a + HCOOH_a \tag{21}$$

$$CH_3CH_2O_a^- \rightarrow HCHO_a \tag{22}$$

$$CH_3O_a \rightarrow HCHO_a + \frac{1}{2}H_{2\,g} \tag{23}$$

$$CH_3OH \rightarrow HCHO_a + H_{2\,ads} \tag{24}$$

It can be seen that routes through Reactions (20) and (21) are not only a possible source of formaldehyde, but also of formic acid.

Some authors working with photocatalytic degradation of acetaldehyde have also observed that the initially formed species, 3-hydroxybutanal and crotonaldehyde, are converted to other intermediates upon illumination of the catalyst. Among the main intermediates, they have identified formic acid, acetic acid and formaldehyde, and this is in accordance with our results [40,64].

It is important to bear in mind that some of the species identified on the catalyst surface after the PDC process can also be formed directly in the gas phase. Indeed, gas-phase acetaldehyde decomposition by NTP can lead to the formation of methoxy radical, as well as acetone, acetic acid and formaldehyde [78], which can subsequently adsorb on the surface.

3.3. Proposed Simplified Mechanism for Acetaldehyde Decomposition Using the Plasma/Catalysis Process

From the identified gaseous by-products, surface IRbands' assignment and proposed adsorption/ degradation routes—a schematic diagram of the possible pathways for acetaldehyde degradation over $Ag/TiO_2/SiO_2$ in the PDC process is presented in Figure 4. It is worth noting that we have not detected formaldehyde at the exit of the DBD reactor whereas it has been detected on the catalyst surface. There are two reasons for this: formaldehyde is a short-lived species [77] and it is strongly adsorbed onto silver catalyst [79].

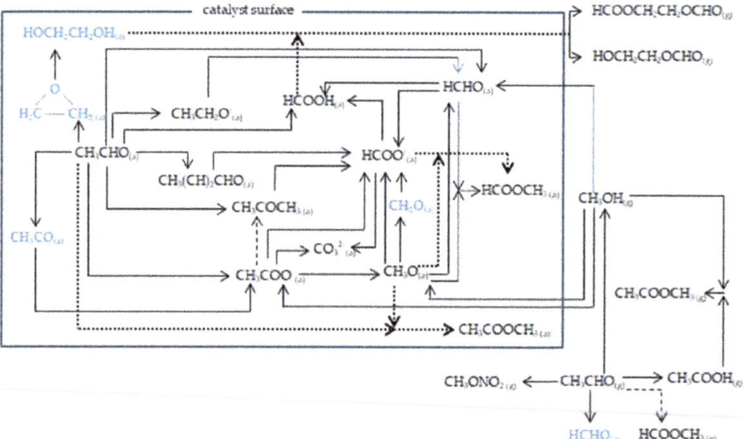

Figure 4. Schematic diagram of the proposed reaction pathway of the PDC degradation of acetaldehyde on Ag/TiO$_2$/SiO$_2$. The line patterns (solid line, dotted line, dashed line) represent, respectively, either direct routes or routes involving reactions with simple radicals/atoms/sites on the surface (H/OH/O/V$_O$/S), routes involving two surface species and routes leading to the formation of by-products observed only at high energy. Blue color compounds: not detected/identified.

It is very likely that EdF and EmF are formed through the esterification of ethylene glycol (EG) and formic acid, even if there is no evidence of the presence of EG either on the effluent or on the surface. However, it is well-known that silver catalysts are commonly used in the manufacturing process for the direct oxidation of ethylene to ethylene oxide (EO) [80], and EO is a precursor of EG [81,82]. Besides, ethylene and EO can be formed from acetaldehyde decomposition on the surface [83–86]. Thus, EdF and EmF formation through an EG pathway is very plausible in the PDC process. In addition, low-energy, electron-induced processes on Me have been proven to induce the formation of methoxy (CH$_3$O) and hydroxymethyl (CH$_2$OH) radicals via electron impact excitation, which, followed by radical–radical coupling, can lead to EG and formic acid formation [87], therefore explaining EmF and EdF production with NTP alone.

In this diagram we did not include the pathways leading to CO$_2$ and CO formation as there are many possibilities. The different routes to CO$_x$ can involve many of the different adsorbed species or gaseous organic compounds. Nevertheless, it is worth noting that increasing the energy in the PDC process also increases the selectivity towards CO. In order to explain this result, we have proposed in a previous work a decarbonylation process of acetaldehyde on the silver-based catalyst [27]. However, this higher CO selectivity can also be the result of the by-products' decomposition pathways. According to a study on the mechanism of the heterogeneously-catalyzed oxidation of organic molecules on metal oxides, it has been stated that formate ions rather easily decompose to CO over metal oxides [88]. Furthermore, an experimental study of ozone catalytic oxidation of gaseous formaldehyde using a TiO$_2$-supported catalyst at room temperature has shown that CO$_x$ selectivity is dependent on the O$_3$:HCHO ratio and the relative humidity (RH). In this work, for a O$_3$:HCHO ratio of 5 (best ratio) and 20% RH, 90% of HCHO removal and a CO/CO$_2$ ratio of 1.5 were achieved, making CO the major product [89]. Similar results regarding higher conversion towards CO than CO$_2$ where obtained when using TiO$_2$-supported catalysts for ethanol oxidation; and it is well-known that acetaldehyde is one of the major intermediate species in ethanol decomposition [70].

As presented in Section 2.1, except from Ace and the fraction corresponding to the unknown compounds, the same by-products are obtained with the NTP and PDC processes. This fact allows the drawing of an initial conclusion concerning Ace and MeF formation. Indeed, at the experimental conditions used, Ace seems to be generated only on the surface, and most certainly through the ketonization of acetic acid/acetate species. In addition, with NTP at 150 $J \cdot L^{-1}$ there is no formation of MeF, whereas with the PDC process it is already formed at 70 $J \cdot L^{-1}$, and in equivalent amounts to the other acetate/formate compounds. Thus, it is clear that the catalyst surface promotes MeF formation. As mentioned earlier, Me can be activated by surface oxygen. In a study on methanol oxidation over silver catalysts, the authors reported that the catalytic oxidative reaction of methanol to methyl formate is related to a synergic process concerning oxygen species on the silver surface. The mechanism proposed in this O_s-Me system starts with methanol dehydrogenation to form adsorbed formaldehyde and methoxy; thereafter, formaldehyde reacts with adsorbed OH or O to form adsorbed formate, that reacts with methoxy and forms methyl formate [79,90]. Thus, Me formed as a by-product of acetaldehyde decomposition can be the precursor of MeF, with formaldehyde and CH_3O as intermediates. In addition, as presented in Figure 1, NTP at 150 $J \cdot L^{-1}$ produces more Me and AcA than the PDC process. Thus, we can suggest that the lower amount of Me is related to its decomposition into formaldehyde and CH_3O, that will in turn produce MeF; and the lower amount of AcA is probably related to the preferential decomposition of acetate through other routes than the one at its origin. From spectra in Figure 3 we have seen that bands corresponding to methoxy, acetate and formaldehyde are present at 70 $J \cdot L^{-1}$, which stresses the feasibility of the proposed mechanisms for Ace and MeF formation.

Increasing SIE from 70 to 350 $J \cdot L^{-1}$ increases acetaldehyde degradation from 33% to 87%, which leads to an important rise in the concentration of all the identified by-products. CO concentration increases 8-fold, whereas the other compounds experience a 2.5- to 4-fold increase. This result confirms the fact that acetaldehyde degradation follows a pathway favoring CO formation. Concerning the catalyst surface composition, in Figure 3b we can see the emergence of a peak at 2914 cm^{-1}, which has been assigned in Section 2.3 either to adsorbed acetaldehyde or to an intermediate formed from formaldehyde oxidation or disproportionation. Considering all the above, this peak could rather be assigned to the intermediate formed from formaldehyde oxidation to formate species, that will subsequently lead to the formation of the different formate compounds. This is supported by the loss of the band centered at 1870 cm^{-1}, attributed to formate species, for which desorption from the surface may be favored at this energy. Similarly, the growth of a rather broad band in the 1300–950 cm^{-1} region observed in Figure 3a should rather be assigned to methoxy species that can also promote formate species formation.

When increasing SIE to 1150 $J \cdot L^{-1}$, two different behaviors among the by-products were observed: (i) those whose concentration remains rather constant (CO_2, MeO, MeF and Ace), (ii) those whose concentration almost doubled (CO, MeA, EdF, EmF and AcA). Therefore, it seems that once formate/formaldehyde are formed, the path that would be favored is the one towards formic acid and then EmF and EdF. Additionally, acetates will be transformed into AcA and MeA rather than into Ace.

4. Materials and Methods

4.1. Catalyst Preparation

Silver nanoparticles deposition was performed following the method presented in a previous study [91]. In this same study, atomic force microscopy (AFM, Veeco, Munich, Germany), scanning electron microscopy (SEM, Supra 40 VP, Zeiss, France) and transmission electron microscopy (TEM, 200 kV JEM 2011, JEOL, France) measurements showing a homogeneous dispersion of silver nanoclusters on the TiO_2 monolayer can be found. The silver nanoparticles were deposited on 150-μm-diameter SiO_2 pellets (Sigma-Aldrich silica gel, St. Louis, MO, USA). The pellets were first covered by a size-selected titanium oxo-alkoxy (TOA, Sigma-Aldrich, St. Louis, MO, USA, 99.999% purity)) nanoparticles monolayer synthesized in a rapid micromixing sol-gel reactor. Then, the silver nanoclusters (10 nm mean diameter) were grown on the TOA monolayer by photocatalytic reduction of an Ag^+ aqueous solution ($AgNO_3$, Prolabo, Prolabo, Paris, France, 99.8% purity), under ultraviolet (UV) irradiation (Philips, Haarlem, The Netherlands, operating at 362 ± 10 nm, 8W). Finally, the pellets were washed in distilled water and dried at 80 °C for 4 h. A 303 m^2g^{-1} BET (Brunauer–Emmet–Teller) (Beckman Coulter, Brea, CA, USA, SA^{TM} 3100) specific surface area was measured.

4.2. Plasma/Catalysis Process

A DBD reactor previously described [27], was used for acetaldehyde removal at room temperature and atmospheric pressure. The DBD reactor is composed of a quartz cylinder externally covered by a grounded stainless steel. A sinusoidal high voltage is applied to a stainless steel cylindrical center electrode through a 5-mm gap. This high voltage is obtained using a voltage amplifier (TREK, 20/20 C, France), coupled to a frequency generator. Different values of the specific input energy (SIE in $J \cdot L^{-1}$), defined as the ratio of the discharge power (W) over the flow rate ($L \cdot s^{-1}$), were obtained by changing the frequency of the signal. When using the PDC process, the $Ag/TiO_2/SiO_2$ nanostructured catalyst was placed in the plasma discharge zone.

The pollutant mixture was composed of acetaldehyde (1000 ppmC), oxygen (20%) and nitrogen (balance) (Air Liquide Alphagaz^{TM} 1), and the flow rate was set to 100 $mL \cdot min^{-1}$ (Bronkhorst El-Flow® Select). The gas was admitted at the bottom of the DBD reactor and flowed upward through the fluidized bed. At the exit, it was sent to a set of analyzers: gas chromatography (Shimadzu GC-2110, Shimadzu, Noisiel, France), for quantitative analysis of residual acetaldehyde; gas chromatography coupled to mass spectrometry, for identification of the gaseous by-products (Shimadzu 2110-GC coupled to a QP2010S-MS, Shimadzu, Noisiel, France); and an infrared multigas analyzer (MIR 9000, Environment SA, France), for CO_2 and CO quantification. The plasma-catalysis experiments were carried out in two stages: (i) the catalyst surface was saturated by acetaldehyde at room temperature and atmospheric pressure, and (ii) the plasma discharge was switched on for 30 min to induce acetaldehyde removal on the surface in the same temperature and pressure conditions. The surface was then characterized by DRIFTS analysis.

4.3. DRIFTS Measurements

DRIFTS analyses were performed with a Shimadzu IRPrestige-21 spectrophotometer (DLATGS detector, 100 scans accumulation, and 4 cm^{-1} resolution, Shimadzu, France). A Pike DiffusIRTM accessory (Pike, Eurolabo, France) containing a chamber equipped with a KBr window was adapted on the FTIR spectrophotometer. This chamber can contain a cup with approximately 10 mg of catalyst. All DRIFTS spectra presented were acquired in transmittance mode.

5. Conclusions

In summary, we used DRIFT spectroscopy to identify the adsorbed species formed on an $Ag/TiO_2/SiO_2$ surface during acetaldehyde adsorption and degradation with a PDC process. The measurements showed that different species such as acetate, formate, methoxy, ethoxy and formaldehyde are present on the surface, once it has been in contact with the plasma.

A pathway that considers the formation of these intermediates via the interaction of adsorbed acetaldehyde, and the plasma-generated species and/or by-products of acetaldehyde decomposition in the gas phase is proposed. The main carbonaceous intermediates identified allowed more detailed mechanism for acetaldehyde degradation to be given, as well as the formation of most of the identified by-products detected in the gas phase leaving the reactor to be explained.

Further investigations are needed to include nitrogenous species detected in the gas phase in the mechanism of acetaldehyde degradation with the PDC process. It is very likely that in CH_3CHO degradation the presence of methanol, one of the intermediate by-products in CH_3CHO oxidation, combined with oxygen activation by silver atoms on the surface, are key elements in the performance of the PDC process.

Author Contributions: Conceptualization, A.V.-G., X.D. and S.S.; Formal analysis, A.V.-G., X.D. and S.S.; Funding acquisition, X.D.; Investigation, A.V.-G., X.D. and S.S.; Methodology, A.V.-G., X.D. and S.S.; Supervision, A.V.-G. and X.D.; Validation, A.V.-G. and X.D.; Writing—original draft, A.V.-G. All authors have read and agreed to the published version of the manuscript.

Funding: This research was funded by the ANR Programme Blanc grant ANR-08-BLAN-015–01 à 04)) and Region Ile de France grant IF-110005975 and 10005961).

Acknowledgments: The authors wish to thank Nicolas Fagnon for the Labview program used for the automated data acquisition during the DRIFTS experiments.

Conflicts of Interest: The authors declare no conflict of interest.

References

1. Li, K.; Ji, J.; Huang, H.; He, M. Efficient activation of Pd/CeO$_2$ catalyst by non-thermal plasma for complete oxidation of indoor formaldehyde at room temperature. *Chemosphere* **2020**, *246*, 125762. [CrossRef] [PubMed]

2. Lee, H.; Song, M.Y.; Ryu, S.; Park, Y.-K. Acetaldehyde oxidation under high humidity using a catalytic non-thermal plasma system over Mn-loaded Y zeolites. *Mater. Lett.* **2020**, *262*, 127051. [CrossRef]

3. Vandenbroucke, A.M.; Nguyen Dinh, M.T.; Nuns, N.; Giraudon, J.M.; De Geyter, N.; Leys, C.; Lamonier, J.F.; Morent, R. Combination of non-thermal plasma and Pd/LaMnO$_3$ for dilute trichloroethylene abatement. *Chem. Eng. J.* **2016**, *283*, 668–675. [CrossRef]

4. Magureanu, M.; Mandache, N.B.; Eloy, P.; Gaigneaux, E.M.; Parvulescu, V.I. Plasma-assisted catalysis for volatile organic compounds abatement. *Appl. Catal. B Environ.* **2005**, *61*, 12–20. [CrossRef]

5. Vandenbroucke, A.M.; Morent, R.; De Geyter, N.; Leys, C. Non-thermal plasmas for non-catalytic and catalytic VOC abatement. *J. Hazard. Mater.* **2011**, *195*, 30–54. [CrossRef]

6. Fan, X.; Zhu, T.L.; Wang, M.Y.; Li, X.M. Removal of low-concentration BTX in air using a combined plasma catalysis system. *Chemosphere* **2009**, *75*, 1301–1306. [CrossRef]

7. Feng, X.; Liu, H.; He, C.; Shen, Z.; Wang, T. Synergistic effects and mechanism of a non-thermal plasma catalysis system in volatile organic compound removal: A review. *Catal. Sci. Technol.* **2018**, *8*, 936–954. [CrossRef]

8. Wang, B.; Xu, X.; Xu, W.; Wang, N.; Xiao, H.; Sun, Y.; Huang, H.; Yu, L.; Fu, M.; Wu, J.; et al. The Mechanism of Non-thermal Plasma Catalysis on Volatile Organic Compounds Removal. *Catal. Surv. Asia* **2018**, *22*, 73–94. [CrossRef]

9. Schiavon, M.; Torretta, V.; Casazza, A.; Ragazzi, M. Non-thermal Plasma as an Innovative Option for the Abatement of Volatile Organic Compounds: A Review. *Water Air Soil Pollut.* **2017**, *228*, 388. [CrossRef]

10. Perillo, R.; Ferracin, E.; Giardina, A.; Marotta, E.; Paradisi, C. Efficiency, products and mechanisms of ethyl acetate oxidative degradation in air non-thermal plasma. *J. Phys. D Appl. Phys.* **2019**, *52*, 295206. [CrossRef]

11. Bal, K.M.; Huygh, S.; Bogaerts, A.; Neyts, E.C. Effect of plasma-induced surface charging on catalytic processes: Application to CO$_2$ activation. *Plasma Sources Sci. Technol.* **2018**, *27*, 024001. [CrossRef]

12. Song, H.; Hu, F.; Peng, Y.; Li, K.; Bai, S.; Li, J. Non-thermal plasma catalysis for chlorobenzene removal over CoMn/TiO$_2$ and CeMn/TiO$_2$: Synergistic effect of chemical catalysis and dielectric constant. *Chem. Eng. J.* **2018**, *347*, 447–454. [CrossRef]

13. Barakat, C.; Gravejat, P.; Guaitella, O.; Thevenet, F.; Rousseau, A. Oxidation of isopropanol and acetone adsorbed on TiO$_2$ under plasma generated ozone flow: Gas phase and adsorbed species monitoring. *Appl. Catal. B Environ.* **2014**, *147*, 302–313. [CrossRef]

14. Xu, W.; Xu, X.; Wu, J.; Fu, M.; Chen, L.; Wang, N.; Xiao, H.; Chen, X.; Ye, D. Removal of toluene in adsorption-discharge plasma systems over a nickel modified SBA-15 catalyst. *RSC Adv.* **2016**, *6*, 104104–104111. [CrossRef]

15. Sultana, S.; Vandenbroucke, A.M.; Mora, M.; Jiménez-Sanchidrián, C.; Romero-Salguero, F.J.; Leys, C.; De Geyter, N.; Morent, R. Post plasma-catalysis for trichloroethylene decomposition over CeO$_2$ catalyst: Synergistic effect and stability test. *Appl. Catal. B Environ.* **2019**, *253*, 49–59. [CrossRef]

16. Rivallan, M.; Fourre, E.; Aiello, S.; Tatibouet, J.M.; Thibault-Starzyk, F. Insights into the Mechanisms of Isopropanol Conversion on γ-Al$_2$O$_3$ by Dielectric Barrier Discharge. *Plasma Process. Polym.* **2012**, *9*, 850–854. [CrossRef]

17. Stere, C.E.; Adress, W.; Burch, R.; Chansai, S.; Goguet, A.; Graham, W.G.; Hardacre, C. Probing a Non-Thermal Plasma Activated Heterogeneously Catalyzed Reaction Using in Situ DRIFTS-MS. *ACS Catal.* **2015**, *5*, 956–964. [CrossRef]

18. Manabe, R.; Okada, S.; Inagaki, R.; Oshima, K.; Ogo, S.; Sekine, Y. Surface Protonics Promotes Catalysis. *Sci. Rep.* **2016**, *6*, 38007. [CrossRef]

19. Sun, Y.; Li, J.; Chen, P.; Wang, B.; Wu, J.; Fu, M.; Chen, L.; Ye, D. Reverse water-gas shift in a packed bed DBD reactor: Investigation of metal-support interface towards a better understanding of plasma catalysis. *Appl. Catal. A Gen.* **2020**, *591*, 117407. [CrossRef]

20. Xu, S.; Chansai, S.; Stere, C.; Inceesungvorn, B.; Goguet, A.; Wangkawong, K.; Taylor, S.F.R.; Al-Janabi, N.; Hardacre, C.; Martin, P.A.; et al. Sustaining metal–organic frameworks for water–gas shift catalysis by non-thermal plasma. *Nat. Catal.* **2019**, *2*, 142–148. [CrossRef]

21. Xu, S.; Chansai, S.; Shao, Y.; Xu, S.; Wang, Y.-c.; Haigh, S.; Mu, Y.; Jiao, Y.; Stere, C.E.; Chen, H.; et al. Mechanistic study of non-thermal plasma assisted CO_2 hydrogenation over Ru supported on MgAl layered double hydroxide. *Appl. Catal. B Environ.* **2020**, *268*, 118752. [CrossRef]

22. Kim, H.-H.; Ogata, A.; Schiorlin, M.; Marotta, E.; Paradisi, C. Oxygen Isotope ($^{18}O_2$) Evidence on the Role of Oxygen in the Plasma-Driven Catalysis of VOC Oxidation. *Catal. Lett.* **2010**, *141*, 277–282. [CrossRef]

23. Christensen, P.A.; Mashhadani, Z.T.A.W.; Md Ali, A.H.B.; Manning, D.A.C.; Carroll, M.A.; Martin, P.A. An in situ FTIR study of the plasma- and thermally-driven reaction of isopropyl alcohol at CeO_2: Evidence for a loose transition state involving Ce^{3+}? *PCCP* **2019**, *21*, 1354–1366. [CrossRef] [PubMed]

24. Rodrigues, A.; Tatibouët, J.-M.; Fourré, E. Operando DRIFT Spectroscopy Characterization of Intermediate Species on Catalysts Surface in VOC Removal from Air by Non-thermal Plasma Assisted Catalysis. *Plasma Chem. Plasma Process.* **2016**, *36*, 901–915. [CrossRef]

25. Jia, Z.; Wang, X.; Thevenet, F.; Rousseau, A. Dynamic probing of plasma-catalytic surface processes: Oxidation of toluene on CeO_2. *Plasma Process. Polym.* **2017**, *14*, 1600114. [CrossRef]

26. Jia, Z.; Rousseau, A. Sorbent track: Quantitative monitoring of adsorbed VOCs under in-situ plasma exposure. *Sci. Rep.* **2016**, *6*, 31888. [CrossRef]

27. Jia, Z.; Vega-González, A.; Amar, M.B.; Hassouni, K.; Tieng, S.; Touchard, S.; Kanaev, A.; Duten, X. Acetaldehyde removal using a diphasic process coupling a silver-based nano-structured catalyst and a plasma at atmospheric pressure. *Catal. Today* **2013**, *208*, 82–89. [CrossRef]

28. Sauce, S.; Vega-González, A.; Jia, Z.; Touchard, S.; Hassouni, K.; Kanaev, A.; Duten, X. New insights in understanding plasma-catalysis reaction pathways: Study of the catalytic ozonation of an acetaldehyde saturated $Ag/TiO_2/SiO_2$ catalyst. *Eur. Phys. J. Appl. Phys.* **2015**, *71*, 20805. [CrossRef]

29. Klett, C.; Duten, X.; Tieng, S.; Touchard, S.; Jestin, P.; Hassouni, K.; Vega-González, A. Acetaldehyde removal using an atmospheric non-thermal plasma combined with a packed bed: Role of the adsorption process. *J. Hazard. Mater.* **2014**, *279C*, 356–364. [CrossRef]

30. Mann, A.K.P.; Wu, Z.; Calaza, F.C.; Overbury, S.H. Adsorption and Reaction of Acetaldehyde on Shape-Controlled CeO_2 Nanocrystals: Elucidation of Structure–Function Relationships. *ACS Catal.* **2014**, *4*, 2437–2448. [CrossRef]

31. Singh, M.; Zhou, N.; Paul, D.K.; Klabunde, K.J. IR spectral evidence of aldol condensation: Acetaldehyde adsorption over TiO_2 surface. *J. Catal.* **2008**, *260*, 371–379. [CrossRef]

32. Idriss, H.; Diagne, C.; Hindermann, J.P.; Kiennemann, A.; Barteau, M.A. Reactions of Acetaldehyde on CeO_2 and CeO_2-supported catalysts. *J. Catal.* **1995**, *155*, 219–237. [CrossRef]

33. Rekoske, J.E.; Barteau, M.A. Competition between Acetaldehyde and Crotonaldehyde during Adsorption and Reaction on Anatase and Rutile Titanium Dioxide. *Langmuir ACS J. Surf. Colloids* **1999**, *15*, 11. [CrossRef]

34. Ordomsky, V.V.; Sushkevich, V.L.; Ivanova, I.I. Study of acetaldehyde condensation chemistry over magnesia and zirconia supported on silica. *J. Mol. Catal. A Chem.* **2010**, *333*, 85–93. [CrossRef]

35. Natal Santiago, M.A.; Hill, J.S.; Dumesic, J.A. Studies of the adsorption of acetaldehyde, methyl acetate, ethyl acetate, and methyl trifluoroacetate on silica. *J. Mol. Catal. A Chem.* **1999**, *140*, 16. [CrossRef]

36. Finkelstein-Shapiro, D.; Buchbinder, A.M.; Vijayan, B.; Bhattacharyya, K.; Weitz, E.; Geiger, F.M.; Gray, K.A. Identification of Binding Sites for Acetaldehyde Adsorption on Titania Nanorod Surfaces Using CIMS. *Langmuir ACS J. Surf. Colloids* **2011**, *27*, 14842–14848. [CrossRef] [PubMed]

37. Raskó, J.; Kiss, J. Adsorption and surface reactions of acetaldehyde on alumina-supported noble metal catalysts. *Catal. Lett.* **2005**, *101*, 71–77. [CrossRef]

38. Chang, C.-A.; Ray, B.; Paul, D.K.; Demydov, D.; Klabunde, K.J. Photocatalytic reaction of acetaldehyde over SrTiO$_3$ nanoparticles. *J. Mol. Catal. A Chem.* **2008**, *281*, 99–106. [CrossRef]

39. Young, Z.D.; Hanspal, S.; Davis, R.J. Aldol Condensation of Acetaldehyde over Titania, Hydroxyapatite, and Magnesia. *ACS Catal.* **2016**, *6*, 3193–3202. [CrossRef]

40. Hauchecorne, B.; Terrens, D.; Verbruggen, S.; Martens, J.A.; Van Langenhove, H.; Demeestere, K.; Lenaerts, S. Elucidating the photocatalytic degradation pathway of acetaldehyde: An FTIR in situ study under atmospheric conditions. *Appl. Catal. B Environ.* **2011**, *106*, 630–638. [CrossRef]

41. Zaki, M.I.; Hasan, M.A.; Pasupulety, L. Surface Reactions of Acetone on Al$_2$O$_3$, TiO$_2$, ZrO$_2$, and CeO$_2$: IR Spectroscopic Assessment of Impacts of the Surface Acid–Base Properties. *Langmuir ACS J. Surf. Colloids* **2001**, *17*, 768–774. [CrossRef]

42. Wang, W.; Xu, D.; Cheng, B.; Yu, J.; Jiang, C. Hybrid carbon@TiO$_2$ hollow spheres with enhanced photocatalytic CO$_2$ reduction activity. *J. Mater. Chem. A* **2017**, *5*, 5020–5029. [CrossRef]

43. Kasimayan, U.; Nadarajan, A.; Singaravelu, C.M.; Pan, G.T.; Kandasamy, J.; Yang, T.C.; Lin, J.H. In-situ DRIFT investigation of photocatalytic reduction and oxidation properties of SiO$_2$@alpha-Fe$_2$O$_3$ core-shell decorated RGO nanocomposite. *Sci. Rep.* **2020**, *10*, 2128. [CrossRef] [PubMed]

44. Yu, Z.; Chuang, S. In situ IR study of adsorbed species and photogenerated electrons during photocatalytic oxidation of ethanol on TiO$_2$. *J. Catal.* **2007**, *246*, 118–126. [CrossRef]

45. Lukaski, A.C.; Muggli, D.S. Photocatalytic oxidation of methyl formate on TiO$_2$: A transient DRIFTS study. *J. Catal.* **2004**, *223*, 250–261. [CrossRef]

46. Backes, M.J.; Lukaski, A.C.; Muggli, D.S. Active sites and effects of H$_2$O and temperature on the photocatalytic oxidation of ^{13}C-acetic acid on TiO$_2$. *Appl. Catal. B Environ.* **2005**, *61*, 21–35. [CrossRef]

47. Chuang, C.-C.; Wu, W.-C.; Huang, M.-C.; Huang, I.-C.; Lin, J.-L. FTIR Study of Adsorption and Reactions of Methyl Formate on Powdered TiO$_2$. *J. Catal.* **1999**, *185*, 12. [CrossRef]

48. Coronado, J.M.; Kataoka, S.; Tejedor-Tejedor, I.; Anderson, M.A. Dynamic phenomena during the photocatalytic oxidation of ethanol and acetone over nanocrystalline TiO$_2$: Simultaneous FTIR analysis of gas and surface species. *J. Catal.* **2003**, *219*, 219–230. [CrossRef]

49. Gazsi, A.; Koós, A.; Bánsági, T.; Solymosi, F. Adsorption and decomposition of ethanol on supported Au catalysts. *Catal. Today* **2011**, *160*, 70–78. [CrossRef]

50. Liao, L.-F.; Lien, C.-F.; Lin, J.-L. FTIR study of adsorption and photoreactions of acetic acid on TiO$_2$. *Phys. Chem. Chem. Phys.* **2001**, *3*, 7. [CrossRef]

51. Bratož, S.; Hadži, D.; Sheppard, N. The infra-red absorption bands associated with the COOH and COOD groups in dimeric carboxylic acid—II: The region from 3700 to 1500 cm^{-1}. *Spectrochim. Acta* **1956**, *8*, 249–261. [CrossRef]

52. Raskó, J. Adsorption and reaction of formaldehyde on TiO$_2$-supported Rh catalysts studied by FTIR and mass spectrometry. *J. Catal.* **2004**, *226*, 183–191. [CrossRef]

53. Idriss, H.; Kim, K.S.; Barteau, M.A. Surface-dependent pathways for formaldehyde oxidation and reduction on TiO$_2$(001). *Surf. Sci.* **1992**, *262*, 113–127. [CrossRef]

54. Tóth, M.; Varga, E.; Oszkó, A.; Baán, K.; Kiss, J.; Erdőhelyi, A. Partial oxidation of ethanol on supported Rh catalysts: Effect of the oxide support. *J. Mol. Catal. A Chem.* **2016**, *411*, 377–387. [CrossRef]

55. Hernández-Alonso, M.D.; Tejedor-Tejedor, I.; Coronado, J.M.; Anderson, M.A.; Soria, J. Operando FTIR study of the photocatalytic oxidation of acetone in air over TiO$_2$–ZrO$_2$ thin films. *Catal. Today* **2009**, *143*, 364–373. [CrossRef]

56. Silva, A.; Barandas, A.; Costa, L.; Borges, L.; Mattos, L.; Noronha, F. Partial oxidation of ethanol on Ru/Y$_2$O$_3$ and Pd/Y$_2$O$_3$ catalysts for hydrogen production. *Catal. Today* **2007**, *129*, 297–304. [CrossRef]

57. Gazsi, A.; Schubert, G.; Pusztai, P.; Solymosi, F. Photocatalytic decomposition of formic acid and methyl formate on TiO$_2$ doped with N and promoted with Au. Production of H$_2$. *Int. J. Hydrogen Energy* **2013**, *38*, 7756–7766. [CrossRef]

58. Araña, J.; Garriga i Cabo, C.; Doña-Rodríguez, J.M.; González-Díaz, O.; Herrera-Melián, J.A.; Pérez-Peña, J. FTIR study of formic acid interaction with TiO$_2$ and TiO$_2$ doped with Pd and Cu in photocatalytic processes. *Appl. Surf. Sci.* **2004**, *239*, 60–71. [CrossRef]

59. Miller, K.L.; Lee, C.W.; Falconer, J.L.; Medlin, J.W. Effect of water on formic acid photocatalytic decomposition on TiO$_2$ and Pt/TiO$_2$. *J. Catal.* **2010**, *275*, 294–299. [CrossRef]

60. Liao, L.-F.; Wu, W.-C.; Chen, C.-Y.; Lin, J.-L. Photooxidation of Formic Acid vs. Formate and Ethanol vs. Ethoxy on TiO$_2$ and Effect of Adsorbed Water on the Rates of Formate and Formic Acid Photooxidation. *J. Phys. Chem. B* **2001**, *105*, 7678–7685. [CrossRef]

61. Li, C.; Domen, K.; Maruya, K.; Onishi, T. Spectroscopic Identification of Adsorbed Species Derived from Adsorption and Decomposition of Formic Acid, Methanol, and Formaldehyde on Cerium Oxide. *J. Catal.* **1990**, *125*, 11. [CrossRef]

62. Xie, B.; Wong, R.J.; Tan, T.H.; Higham, M.; Gibson, E.K.; Decarolis, D.; Callison, J.; Aguey-Zinsou, K.F.; Bowker, M.; Catlow, C.R.A.; et al. Synergistic ultraviolet and visible light photo-activation enables intensified low-temperature methanol synthesis over copper/zinc oxide/alumina. *Nat. Commun.* **2020**, *11*, 1615. [CrossRef] [PubMed]

63. Sadykov, V.A.; Eremeev, N.F.; Sadovskaya, E.M.; Chesalov, Y.A.; Pavlova, S.N.; Rogov, V.A.; Simonov, M.N.; Bobin, A.S.; Glazneva, T.S.; Smal, E.A.; et al. Detailed Mechanism of Ethanol Transformation into Syngas on Catalysts Based on Mesoporous MgAl$_2$O$_4$ Support Loaded with Ru + Ni/(PrCeZrO or MnCr$_2$O$_4$) Active Components. *Top. Catal.* **2020**, *63*, 166–177. [CrossRef]

64. Topalian, Z.; Stefanov, B.I.; Granqvist, C.G.; Österlund, L. Adsorption and photo-oxidation of acetaldehyde on TiO$_2$ and sulfate-modified TiO$_2$: Studies by in situ FTIR spectroscopy and micro-kinetic modeling. *J. Catal.* **2013**, *307*, 265–274. [CrossRef]

65. Raskó, J.; Kecskés, T.; Kiss, J. FT-IR and mass spectrometric studies on the interaction of acetaldehyde with TiO$_2$-supported noble metal catalysts. *Appl. Catal. A Gen.* **2005**, *287*, 244–251. [CrossRef]

66. Idriss, H.; Kim, K.S.; Barteau, M.A. Carbon-Carbon Bond Formation via Aldolization of Acetaldehyde on Single Crystal and Polycrystalline TiO$_2$ Surfaces. *J. Catal.* **1993**, *139*, 15. [CrossRef]

67. Kim, K.S.; Barteau, M.A. Structure and Composition Requirements for Deoxygenation, Dehydration, and Ketonization Reactions of Carboxylic Acids on TiO$_2$(001) Single-Crystal Surfaces. *J. Catal.* **1990**, *125*, 23. [CrossRef]

68. Sheng, P.Y.; Bowmaker, G.A.; Idriss, H. The Reactions of Ethanol over Au/CeO$_2$. *Appl. Catal. A Gen.* **2004**, *261*, 171–181. [CrossRef]

69. Rorrer, J.E.; Toste, F.D.; Bell, A.T. Mechanism and Kinetics of Isobutene Formation from Ethanol and Acetone over Zn$_x$Zr$_y$O$_z$. *ACS Catal.* **2019**, *9*, 10588–10604. [CrossRef]

70. Velasquez Ochoa, J.; Farci, E.; Cavani, F.; Sinisi, F.; Artiglia, L.; Agnoli, S.; Granozzi, G.; Paganini, M.C.; Malfatti, L. CeO$_x$/TiO$_2$ (Rutile) Nanocomposites for the Low-Temperature Dehydrogenation of Ethanol to Acetaldehyde: A Diffuse Reflectance Infrared Fourier Transform Spectroscopy–Mass Spectrometry Study. *ACS Appl. Nano Mater.* **2019**, *2*, 3434–3443. [CrossRef]

71. Bokhimi, X.; Zanella, R.; Maturano, V.; Morales, A. Nanocrystalline Ag, and Au–Ag alloys supported on titania for CO oxidation reaction. *Mater. Chem. Phys.* **2013**, *138*, 490–499. [CrossRef]

72. Zhao, D.Z.; Shi, C.; Li, X.S.; Zhu, A.M.; Jang, B.W. Enhanced effect of water vapor on complete oxidation of formaldehyde in air with ozone over MnO$_x$ catalysts at room temperature. *J. Hazard. Mater.* **2012**, *239–240*, 362–369. [CrossRef] [PubMed]

73. Nie, L.; Yu, J.; Jaroniec, M.; Tao, F.F. Room-temperature catalytic oxidation of formaldehyde on catalysts. *Catal. Sci. Technol.* **2016**, *6*, 3649–3669. [CrossRef]

74. Sun, Z.; Zhang, X.; Li, H.; Liu, T.; Sang, S.; Chen, S.; Duan, L.; Zeng, L.; Xiang, W.; Gong, J. Chemical looping oxidative steam reforming of methanol: A new pathway for auto-thermal conversion. *Appl. Catal. B Environ.* **2020**, *269*, 118758. [CrossRef]

75. Manzoli, M.; Chiorino, A.; Boccuzzi, F. Decomposition and combined reforming of methanol to hydrogen: A FTIR and QMS study on Cu and Au catalysts supported on ZnO and TiO$_2$. *Appl. Catal. B Environ.* **2005**, *57*, 201–209. [CrossRef]

76. Han, Y.; Liu, C.-J.; Ge, Q. Effect of Pt Clusters on Methanol Adsorption and Dissociation over Perfect and Defective Anatase TiO$_2$(101) Surface. *J. Phys. Chem. C* **2009**, *113*, 20674–20682. [CrossRef]

77. Wittstock, A.; Zielasek, V.; Biener, J.; Friend, C.M.; Bäumer, M. Nanoporous Gold Catalysts for Selective Gas-Phase Oxidative Coupling of Methanol at Low Temperature. *Science* **2010**, *327*, 319. [CrossRef]

78. Klett, C.; Touchard, S.; Vega-González, A.; Redolfi, M.; Bonnin, X.; Hassouni, K.; Duten, X. Experimental and modeling study of the oxidation of acetaldehyde in an atmospheric-pressure pulsed corona discharge. *Plasma Sources Sci. Technol.* **2012**, *21*, 045001. [CrossRef]

79. Yang, Z.; Li, J.; Yang, X.; Wu, Y. Catalytic oxidation of methanol to methyl formate over silver? A new purpose of a traditional catalysis system. *Catal. Lett.* **2005**, *100*, 205–211. [CrossRef]

80. Rebsdat, S.; Mayer, D.; Ethylene Oxide. Ethylene Oxide. In *Ullmann's Encyclopedia of Industrial Chemistry*; Wiley-VCH Verlag GmbH & Co: Weinheim, Germany, 2001. Available online: https://onlinelibrary.wiley.com/doi/abs/10.1002/14356007.a10_117 (accessed on 28 July 2020).

81. Li, Y.; Yue, B.; Yan, S.; Yang, W.; Xie, Z.; Chen, Q.; He, H. Preparation of Ethylene Glycol via Catalytic Hydration with Highly Efficient Supported Niobia Catalyst. *Catal. Lett.* **2004**, *95*, 163–166. [CrossRef]

82. Kandasamy, S.; Samudrala, S.P.; Bhattacharya, S. The route towards sustainable production of ethylene glycol from a renewable resource, biodiesel waste: A review. *Catal. Sci. Technol.* **2019**, *9*, 567–577. [CrossRef]

83. Yu, W.; Mellinger, Z.J.; Barteau, M.A.; Chen, J.G. Comparison of Reaction Pathways of Ethylene Glycol, Acetaldehyde, and Acetic Acid on Tungsten Carbide and Ni-Modified Tungsten Carbide Surfaces. *J. Phys. Chem. C* **2012**, *116*, 5720–5729. [CrossRef]

84. McManus, J.R.; Martono, E.; Vohs, J.M. Selective Deoxygenation of Aldehydes: The Reaction of Acetaldehyde and Glycolaldehyde on Zn/Pt(111) Bimetallic Surfaces. *ACS Catal.* **2013**, *3*, 1739–1750. [CrossRef]

85. Neitzel, A.; Lykhach, Y.; Johánek, V.; Tsud, N.; Skála, T.; Prince, K.C.; Matolín, V.; Libuda, J. Role of Oxygen in Acetic Acid Decomposition on Pt(111). *J. Phys. Chem. C* **2014**, *118*, 14316–14325. [CrossRef]

86. Chang, Y.-C.; Ko, A.-N. Vapor phase reactions of acetaldehyde over type X zeolites. *Appl. Catal. A Gen.* **2000**, *190*, 149–155. [CrossRef]

87. Boamah, M.D.; Sullivan, K.K.; Shulenberger, K.E.; Soe, C.M.; Jacob, L.M.; Yhee, F.C.; Atkinson, K.E.; Boyer, M.C.; Haines, D.R.; Arumainayagam, C.R. Low-energy electron-induced chemistry of condensed methanol: Implications for the interstellar synthesis of prebiotic molecules. *Faraday Discuss.* **2014**, *168*, 249. [CrossRef] [PubMed]

88. Busca, G. Infrared studies of the reactive adsorption of organic molecules over metal oxides and of the mechanisms of their heterogeneously-catalyzed oxidation. *Catal. Today* **1996**, *27*, 40. [CrossRef]

89. Kim, M.; Park, E.; Jurng, J. Oxidation of gaseous formaldehyde with ozone over MnO_x/TiO_2 catalysts at room temperature (25 °C). *Powder Technol.* **2018**, *325*, 368–372. [CrossRef]

90. Wittstock, A.; Biener, J.; Bäumer, M. Nanoporous Gold: A Novel Catalyst with Tunable Properties. *ECS Trans.* **2010**, *28*, 1–13. [CrossRef]

91. Jia, Z.; Ben Amar, M.; Brinza, O.; Astafiev, A.; Nadtochenko, V.; Evlyukhin, A.B.; Chichkov, B.N.; Duten, X.; Kanaev, A. Growth of Silver Nanoclusters on Monolayer Nanoparticulate Titanium-oxo-alkoxy Coatings. *J. Phys. Chem. C* **2012**, *116*, 17239–17247. [CrossRef]

 *catalysts*

Article

Paracetamol Degradation by Catalyst Enhanced Non-Thermal Plasma Process for a Drastic Increase in the Mineralization Rate

Noussaiba Korichi [1], **Olivier Aubry** [1,*], **Hervé Rabat** [1], **Benoît Cagnon** [2] **and Dunpin Hong** [1,*]

[1] GREMI, UMR 7344, CNRS, Université d'Orléans, 45067 Orléans, France; noussaiba.korichi@univ-orleans.fr (N.K.); herve.rabat@univ-orleans.fr (H.R.)

[2] ICMN, UMR 7374, CNRS, Université d'Orléans, 45071 Orléans, France; benoit.cagnon@univ-orleans.fr

* Correspondence: olivier.aubry@univ-orleans.fr (O.A.); dunpin.hong@univ-orleans.fr (D.H.)

Received: 27 July 2020; Accepted: 19 August 2020; Published: 21 August 2020

Abstract: In order to remediate the very poor mineralization of paracetamol in water, even when well degraded by using a Non-Thermal Plasma (NTP) process at a very low dissipated power, a plasma-catalyst coupling process was tested and investigated. A homemade glass fiber supported Fe^{3+} catalyst was immersed in the liquid to be treated in a Dielectric Barrier Discharge plasma reactor. The plasma-catalysis process, at the same low dissipated power, achieved a mineralization rate of 54% with a full conversion rate of paracetamol at 25 mg L^{-1} in initial concentration after 60 min treatment, thanks to Fenton-like effects. The synergetic effects of the plasma-catalysis coupling process also improved the Energy Yield by a factor of two. The catalyst before and after use for treatment was characterized by Brunauer-Emmett-Teller and Thermogravimetric analysis. High-Performance Liquid Chromatography was used to measure the concentration of treated solution and to investigate the intermediates. Two of them, namely 1,4-hydroquinone and 1,4-benzoquinone, were formally identified. Some intermediates are presented in this paper as a function of treatment time and their UV absorbance spectra. NTP processes with and without catalyst coupling were compared in terms of acidity, conductivity, and nitrate concentrations in the treated solution.

Keywords: non-thermal plasma (NTP); heterogeneous Fenton catalyst; advanced oxidation process (AOP); paracetamol degradation; water treatment; mineralization; energy yield

1. Introduction

Currently, significant volumes of water containing pharmaceutical residues are released into the environment, given the number of existing health care establishments or pharmaceutical industries. Pharmaceutical residues in surface waters have negative impacts on planktonic species [1] and some studies indicate that surface water pharmaceutical levels can lead to potential environmental concerns for aquatic ecosystems [2].

To remove pharmaceutical residues from water, various Advanced Oxidation Processes (AOPs) have been studied for many years, using electrochemical methods [3], ozonation [4], UV-light in photocatalytic oxidation [5], Fenton and photo Fenton-like processes [6]. AOP techniques proceed via reactive oxygen species such as ozone [7], hydrogen peroxide [8], or short-lived species such as hydroxyl radicals or atomic oxygen [9].

Non-thermal Plasmas (NTPs) have been tested to treat various organic compounds and drug residues in water and appear to be a relevant AOP technique [10–13]. They have been used in various ways and the efficiencies (conversion, energy yield) were found to depend on the reactor configuration, the nature and concentration of the target molecules, and the water matrix (distilled, pure, or tap waters). Besides the removal of the target compound, it is important to achieve good mineralization, which represents the conversion of the parent molecule into carbon dioxide and water and eventually other inorganic compounds. A low mineralization degree, in spite of high removal efficiency, means that the treated effluent still contains large amounts of organic carbon, i.e., the by-products resulting from the initial compound degradation. These degradation products may sometimes be as harmful or even more dangerous than the pollutant. In practice, complete mineralization is required only for highly toxic compounds. However, high mineralization is usually required in order to make the effluent more amenable to conventional treatment.

Among the papers in the literature on the treatment of pharmaceutical molecules in water using NTPs, only a few articles concern the degradation of Paracetamol molecules [11,13–15]. Panorel et al. [11] used a Pulsed Corona Discharge to oxidize paracetamol and to achieve complete degradation in a short time period with an energy yield in the range of 20–70 g $(kWh)^{-1}$ depending on the power and working gas. Even with a high power, in the range of 200 to 850 W, the Total Organic Carbon (TOC) removal was only about 20% when the paracetamol conversion reached 100%. Baloul et al. [13] used a multiple needle-to-plate Dielectric Barrier Discharge (DBD) reactor for the treatment of paracetamol in aqueous medium with only 0.3 W injected power. In the case of O_2-Ar as injected gas, the conversion rate reached about 100% with an energy yield of around 5 g $(kWh)^{-1}$. The best energy yield of 12 g $(kWh)^{-1}$ was obtained for a gas mixture of air-Ar. Their paper also showed that carboxylic acids and aromatic compounds are the main degradation products of paracetamol in liquid and very low TOC removal rate was obtained, i.e., a very weak mineralization. Recently, Pan and Qiao [14] reported that in their experiment using a cylindrical DBD reactor, the 10 mg L^{-1} paracetamol concentration was removed completely after 5 min treatment with a TOC removal rate of 46.3% after 20 min treatment under 500 W discharge power and 50 mL·min^{-1} air flow rate. Even with this short time of 5 min, the large dissipated power of 500 W implied a rather low energy yield of only 0.36 g $(kWh)^{-1}$. Moreover, one point reported in the paper requires clarification: the total volume was 1500 mL and the recirculating flow rate was 50 mL·min^{-1}, implying a duration of 30 min to ensure that all the liquid has passed through the plasma zone, much longer than the duration of 5 min reported. Furthermore, Iervolino et al. [15] studied the degradation of 25 mg L^{-1} paracetamol in water using a cylindrical DBD reactor. They obtained a complete degradation and mineralization of paracetamol within 15 min using pure O_2 as working gas. However, their best energy yield of 0.59 g $(kWh)^{-1}$ was much lower than that of 12 g $(kWh)^{-1}$ obtained in our previous work [13].

Coupling several AOPs, such as the Fenton process [16,17] and catalytic ozonation [18], which have been extensively studied, may increase the mineralization. Heterogeneous catalytic ozonation has been studied over various materials such as unsupported metal oxides (MnO_2, TiO_2, Al_2O_3), supported metals, or metal oxides (Fe, Mn, Ni, etc.) [7,19]. The catalytic activity is mainly based on the decomposition of ozone at the surface of the catalyst leading to the enhanced generation of HO• radicals. When performing the reaction under oxidative conditions, it was reported that the Fenton process is particularly efficient for the mineralization of a wide range of organic compounds, as the presence of oxygen enhances the degradation of the pharmaceutical residues thanks to the occurrence of auto-oxidation initiated by HO• radicals [20]. Many investigations have focused on the development of stable heterogeneous catalysts to minimize leaching while increasing catalytic activity and long-term stability. However, the development of a more reliable process involving catalysts is still necessary [21].

Recently, studies on the combination of plasma with a catalyst, called the plasma-catalyst process, have been reported for polluted wastewater treatment [22,23]. They showed that the presence of cerium-based catalysts as powder dispersed in the aqueous solution improved the mineralization of phenol. Kusic et al. obtained high phenol removal results by electrical discharge in combination with FeZSM5, which was attributed to the catalytic activity of iron [24].

The present study focused on enhancing the mineralization rate for paracetamol degradation in water using a coupled plasma-catalysis process. For this, NTP-Catalysis coupling based on the Fenton-like mechanism was chosen. In our experimental conditions, NTP alone produced a small amount of $HO^\bullet$ due to the extremely low power dissipated. The coupling with the catalyst allowed for a greater amount of $HO^\bullet$ to be produced thanks to the Fenton-like effect of Fe^{3+}, which regenerates $HO^\bullet$ from H_2O_2. The NTP rector already used in the work of Baloul et al. [13] was used. The manufacturing protocols of the homemade supported catalyst are described in detail in Section 3 together with the experimental setup.

2. Results and Discussion

To demonstrate the decisive role of Plasma-Catalysis coupling in obtaining a good mineralization rate, plasma treatment experiments were conducted with and without coupling to a catalyst for comparison. The experiments were performed under similar conditions to those in our previous work [13]. The treatment was performed with an applied square alternative High Voltage (HV) of ±5.9 kV in amplitude and 500 Hz in frequency. The electrical discharge operated under the streamer regime instead of the spark regime as the latter caused excessive water evaporation and rapid damage to the electrodes [25]. The injected working gas was dry air with a flow rate of 100 standard cubic centimeters per minute (sccm) since it was established in the above-mentioned work that O_2 is mandatory in a DBD discharge for efficient paracetamol degradation.

A homemade supported-catalyst was immersed in the liquid to be treated in the DBD plasma reactor. The advantages of using Fe^{3+} salt in a heterogeneous Fenton (i.e., Fenton-like) process rather than Fe^{2+} (i.e., Fenton) are its lower cost, non-iron-containing sludge generation, and the ease of catalyst separation and reuse [26,27]. In the current work, the Fenton-like process was initiated by the in-situ generation of H_2O_2 by the discharge plasma at an initial pH of 6.1 without adding any external chemical reagent.

2.1. Conversion Rate

In the previous work by Baloul et al. [13], the concentration of paracetamol was measured mainly by using UV absorption at 243 nm. However, some by-products also absorb at this wavelength; hence, the concentration of paracetamol residue was overestimated, as proved by High-Resolution Mass Spectrometry (HRMS) measurements. In the current work, the concentrations were determined from High-Performance Liquid Chromatography (HPLC) chromatograms of the treated solutions and the relative error was estimated to be lower than 1%. For instance, in Figure 1, two chromatograms of the solution after 15 and 45 min' treatment by plasma alone are presented. The peak with a retention time of 4.2 min corresponds to the paracetamol molecule. The area of this peak can be used to determine the concentration of paracetamol through preliminary calibration. The other peaks corresponding to by-products generated from the paracetamol degradation will be discussed later (Section 2.4).

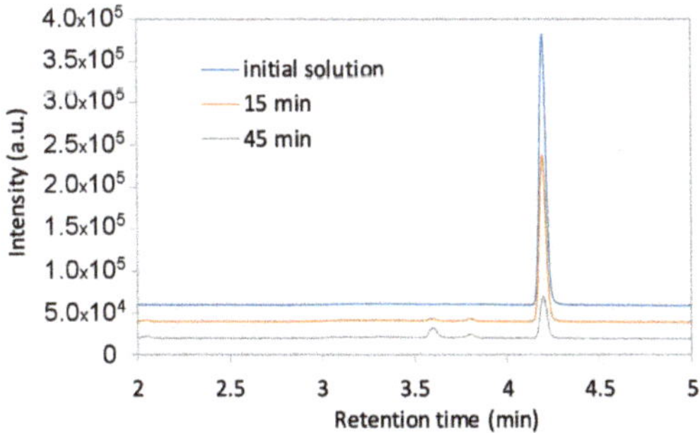

Figure 1. HPLC chromatograms of the solution before and after 15 and 45 min of treatment by NTP alone (U = ±5.9 kV, f = 500 Hz, electrode-liquid-gap = 5 mm, Q_{air} = 100 sccm, [Paracetamol] = 25 mg L^{-1}, V = 40 mL).

The concentrations of paracetamol after plasma treatment with and without catalysis were determined for four treatment durations, namely 15, 30, 45, and 60 min. The calculated conversion rates, τ (%), using these concentrations are presented in Figure 2a. To obtain a complete conversion, a duration of 60 min was necessary. This apparently long treatment time should be regarded in correlation with the low input power (about 0.3 W). Pan and Qiao [14] and Iervolino et al. [15] required a shorter treatment duration, but with a much higher consumed power and lower energy yield.

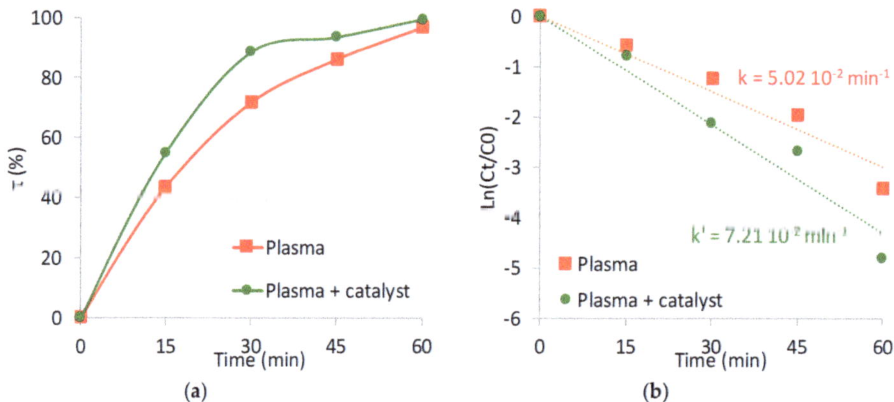

Figure 2. (**a**) Conversion rates (mean values of 3 runs for each operating condition) of paracetamol treatment as a function of treatment time and (**b**) kinetics of paracetamol degradation by NTP without and with catalysis coupling ($U = \pm 5.9$ kV, $f = 500$ Hz, electrode-liquid-gap = 5 mm, $Q_{air} = 100$ sccm, [Paracetamol] = 25 mg L^{-1}, V = 40 mL).

Figure 2a also shows that there is an increase in the conversion rate of paracetamol with the presence of the catalyst compared to the process with plasma alone. This corresponds to a rise of about 40% of the rate constant of the paracetamol degradation with the plasma-catalysis coupling process compared with the treatment by plasma alone, as shown in Figure 2b.

Before evaluating the exact role of the plasma/heterogeneous Fenton coupling on paracetamol degradation, the effects of the catalytic activity of the Fe $(NO_3)_3$ precursor alone in the removal of paracetamol were observed. Neither paracetamol degradation nor mineralization was observed even for a long contact time of 60 min with the catalyst alone. It can therefore be concluded that the contribution of the catalyst without NTP to paracetamol degradation is negligible.

The enhancement in conversion rates can be ascribed to the presence of Fe^{3+} which drives the Fenton reaction to the formation of hydroxyl radicals, considered as the highest reactive radical, via the H_2O_2 produced by the discharge according to the mechanism below [28,29]. Although other radicals $(HO_2^\bullet)$ are produced, their oxidation potential is smaller than that of $HO^\bullet$ species.

$$Fe^{3+} + H_2O_2 \rightarrow Fe^{2+} + HO_2^\bullet + H^+, \tag{1}$$

$$Fe^{3+} + HO_2^\bullet \rightarrow Fe^{2+} + O_2 + H^+ \tag{2}$$

$$H_2O_2 + Fe^{2+} \rightarrow HO^\bullet + HO^- + Fe^{3+} \tag{3}$$

$$HO^\bullet + RH \rightarrow R^\bullet + H_2O \tag{4}$$

$$HO^\bullet + H_2O_2 \rightarrow H_2O + HO_2^\bullet \tag{5}$$

$$HO^\bullet + Fe^{2+} \rightarrow Fe^{3+} + HO^- \tag{6}$$

Figure 3 illustrates the main reactions of the heterogeneous Fenton-like system in the enhancement of mineralization. It starts with the in situ generation of hydroxyl radicals from iron oxidation by H_2O_2. The $HO^\bullet$ will then react directly not only with the paracetamol in water during plasma/catalysis coupling but also with the generated by-products. Thus, a decrease in the content of organic carbon is expected in the presence of catalyst, since the intermediates can be mineralized into CO_2 and H_2O as final products.

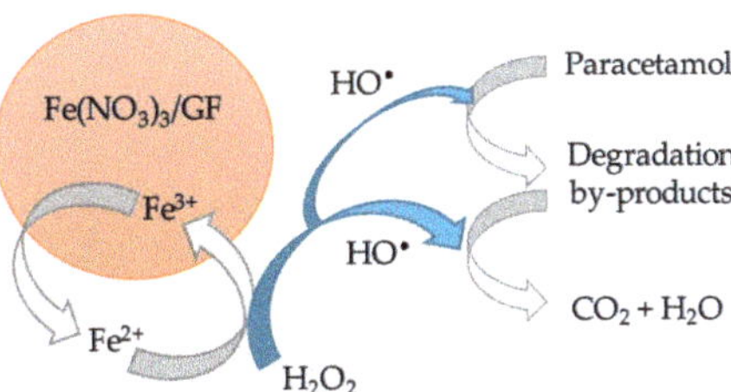

Figure 3. A possible heterogeneous Fenton-like catalytic mechanism in the plasma system.

2.2. Mineralization Rate

The mineralization efficiency (τ_{TOC}) expresses the conversion of the organic carbon into CO_2 and inorganic carbons with the generation of H_2O. Despite the high removal rates of paracetamol by NTP alone, no mineralization has yet been obtained. To understand this "high conversion but low mineralization" phenomenon, Baloul et al. [13] analyzed the treated solution using HRMS and showed that many organic compounds such as carboxylic acids and aromatic compounds were produced from paracetamol degradation. These compounds may be a limiting factor for mineralization of the solution according to Brillas et al. [30].

Figure 4 shows the mineralization rates of solution treated by plasma/catalysis coupling as a function of the treatment time. The τ_{TOC} was lower than the conversion rate of the paracetamol, confirming that the pollutant was transformed upon oxidation into other organic by-products. With the plasma-catalysis coupling, the mineralization rate increased significantly to 30% after 15 min of treatment and reached a maximum of 54% within 60 min. The mineralization rates obtained are comparable to those of Slamani et al. [31], who obtained 60.6% of mineralization after 60 min of paracetamol treatment in water by coupling the Fenton process and a gliding arc discharge, but using much higher input energy than the one consumed by our process.

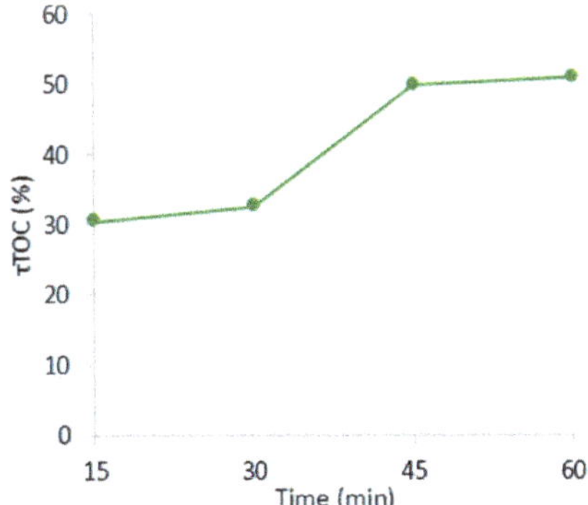

Figure 4. Mineralization rates of paracetamol treated by NTP/Fenton-like coupling (U = ±5.9 kV, f = 500 Hz, electrode-liquid-gap = 5 mm, Q_{air} = 100 sccm, [Paracetamol] = 25 mg L^{-1}, V = 40 mL) (Mineralization rate relative uncertainty <1%).

2.3. Stability and Reuse of the Catalysts

Tests were performed in order to evaluate the stability and the reusability of the catalyst during successive NTP/Fenton-like treatments. Three successive runs of the experiment were carried out using the same working conditions as those mentioned above. As shown in Figure 5, the mineralization efficiency was maintained at a stable level after three successive tests, which proved that the catalyst had good stability and reusability in these operating conditions. Note that the maximum mineralization rate reached 54%.

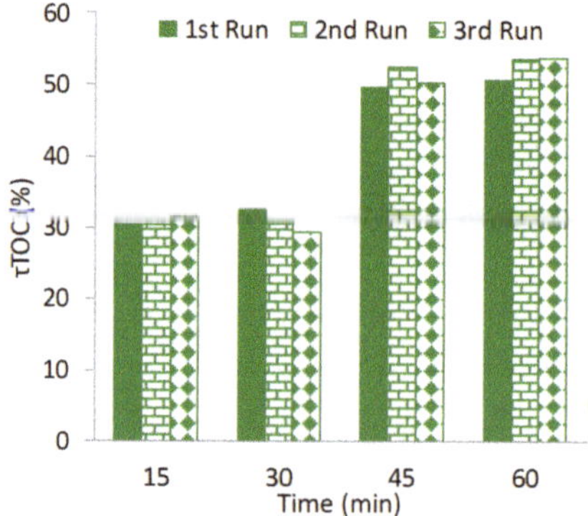

Figure 5. Mineralization rates as a function of the treatment time for each run for paracetamol degradation by NTP/Fenton-like (U = ±5.9 kV, f = 500 Hz, electrode-liquid-gap = 5 mm, Q_{air} = 100 sccm, [Paracetamol] = 25 mg L^{-1}, V = 40 mL).

2.4. HPLC Analyses

Figure 6a displays chromatograms obtained for paracetamol treatments with and without plasma-catalyst coupling for 30 min treatment time. This figure focuses on the species produced due to the paracetamol degradation and corresponding to the various peaks observed. It shows that the plasma/catalysis coupling led to a decrease in the number and area of peaks, which corresponds to an increase in the mineralization rate in this case. In Figure 6b, a zoom of the HPLC chromatograms is displayed, where slight variations in the retention times can be observed. In order to check whether two small shifted peaks corresponded or not to the same molecule, UV-absorption spectra were analyzed, confirming that there were six molecules from 2 to 2.7 min and that different products are formed in the solution treated by the coupled process (for instance, P3 and P5 in Figure 6b). This check was carried out for all the detected peaks.

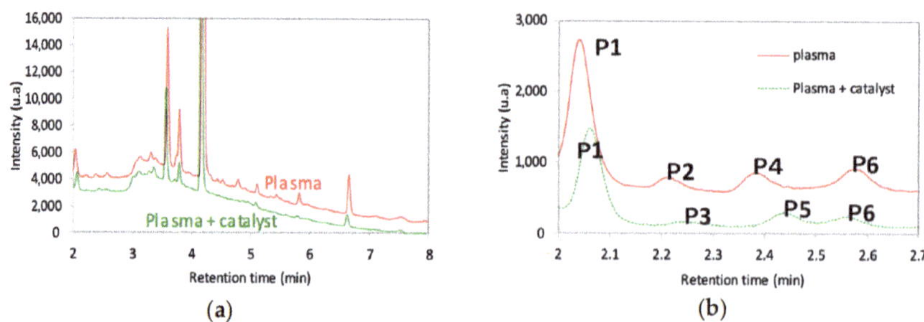

Figure 6. (a) HPLC chromatograms of paracetamol treatments in water at 30 min; (b) zoom of HPLC chromatograms (U = ±5.9 kV, f = 500 Hz, electrode-liquid-gap = 5 mm, Q_{air} = 100 sccm, [Paracetamol] = 25 mg L^{-1}, V = 40 mL).

Figures 7 and 8 show the evolution of the main peaks observed on the chromatograms in terms of peak area and UV absorbance spectra. Variation in the peak area at a given retention time gives information on the evolution of the concentration of a given species. Generally, the degradation products start to be formed soon after the treatment starts, reach a maximum concentration, and then decrease. With plasma/catalyst coupling, it was observed that many of the detected peaks had lower areas than with plasma alone and/or are characterized by a faster decline (Figure 7a,d,f). An exception is the peak at a retention time 2.5 min that only appears for the plasma-catalytic treatment. The peak at 3.4 min is considerably larger in the presence of the catalyst; however, it shows a more pronounced decrease towards long treatment time. The peak at the retention time at 4.8 min was also promoted with the coupled treatment, but only for the longest processing time.

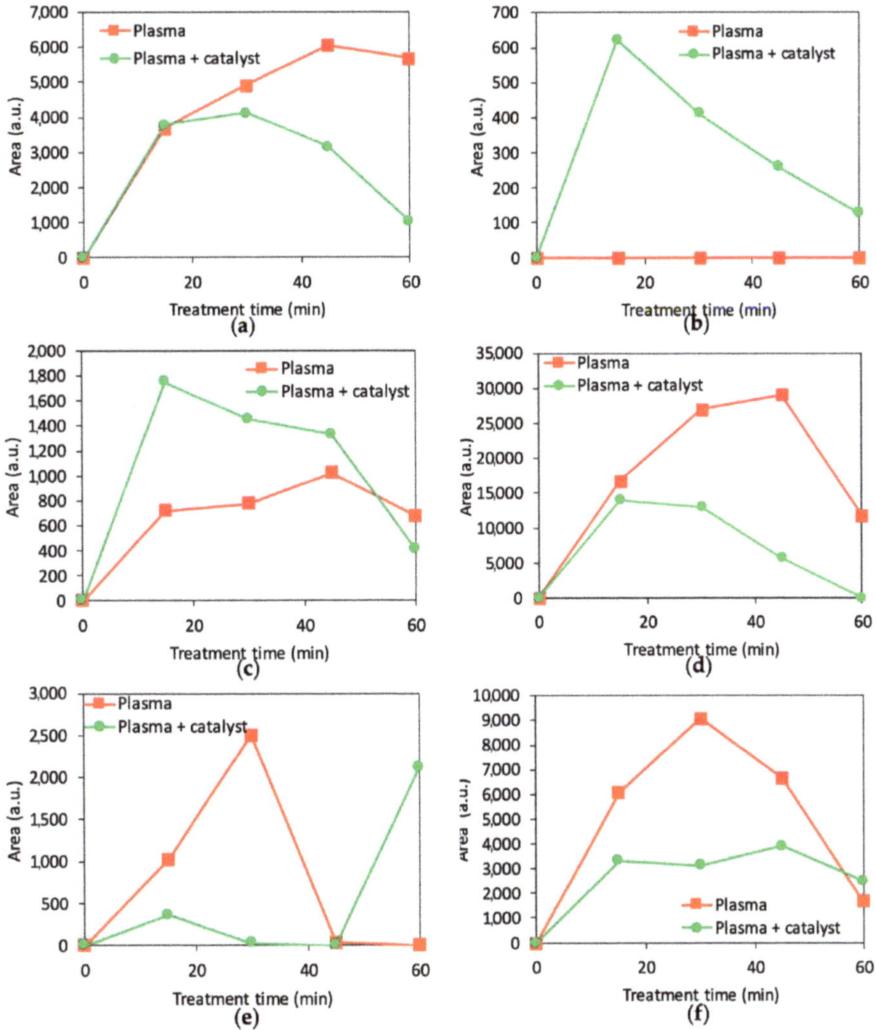

Figure 7. Measured area (mean values of 3 runs for each operating condition) of the main peaks detected by HPLC for the retention times. (**a**) 2.05 min, (**b**) 2.45 min, (**c**) 3.4 min (1,4-hydroquinone), (**d**) 3.6 min, (**e**) 4.8 min (1,4-benzoquinone), (**f**) 6.6 min.

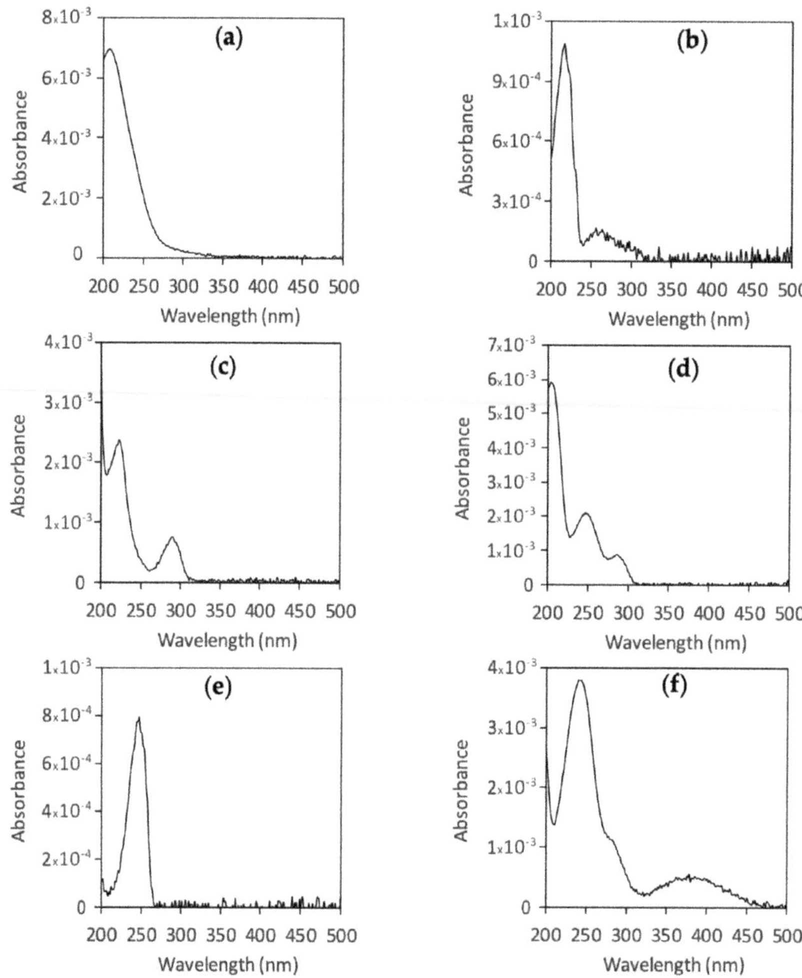

Figure 8. Experimental UV spectra of the main peaks detected by HPLC for the retention times (**a**) 2.05 min, (**b**) 2.45 min, (**c**) 3.4 min (1,4-hydroquinone), (**d**) 3.6 min, (**e**) 4.8 min (1,4-benzoquinone), (**f**) 6.6 min.

Based on the literature and on the information obtained from HPLC (retention times, experimental UV absorbance spectra), the identification of the detected species was attempted. Using commercial standard solutions, only two species were identified by HPLC analyses: 1,4-hydroquinone and 1,4-benzoquinone, which correspond to the peaks detected at 3.4 and 4.8 min. The hydroxylation of aromatic rings is due to the collision reactions of the radicals and other plasma species [32]. The initial attack of HO• on the C-N bond leads to hydroquinone and further oxidation of hydroquinone leads to 1,4-benzoquinone [33,34].

Thanks to standard analyses, the concentration of 1,4-hydroquinone and 1,4-benzoquinone produced from paracetamol treatment can be estimated. For hydroquinone, in plasma-catalyst treatment conditions, the maximum concentration was about 0.34 mg L^{-1} (after 15 min of treatment), while for plasma alone the maximum concentration was lower, with 0.2 mg L^{-1}. For benzoquinone, the maximum concentrations obtained were about 0.02 mg L^{-1} with or without catalysis coupling. The other peaks were not matched with other tested standard species, i.e., nitrophenol and acetamide.

From the results above and the mineralization rates, it can be deduced that most of the species produced are not detected by HPLC since the used column was not appropriate. Baloul et al. [13], for paracetamol treatment by plasma alone, showed by using HRMS that carboxylic acids were produced. These species were no doubt also produced with the plasma-catalyst coupling process to treat paracetamol, but in smaller quantities due to the partial mineralization thanks to the HO$^\bullet$ produced by the Fenton–like effect. The pathway is assumed to be initiated by the attack of HO$^\bullet$ radicals and the substitution of the amide group, followed by the opening of the aromatic ring with the formation of carboxylic acids and finally mineralization. The transformation of primary aromatic products into carboxylic acids [35], such as acetic and formic acids, leads to the production of oxalic acid and then its mineralization by HO$^\bullet$ into CO_2 and H_2O.

2.5. pH and Conductivity of the Treated Solution

Figure 9 shows the pH and conductivity changes of the samples treated by NTP/Fenton-like coupling compared with the ones with plasma alone. The presence of the catalyst leads to less acidification of the samples (i.e., higher pH values, see Figure 9a) and less pronounced increase in the conductivity (Figure 9b) compared to the treatment by plasma alone. For the treatment with plasma alone, the pH of the solution decreases from 6.1 to 3.7 and the conductivity increases from 2.6 μS·cm^{-1} to 122 μS·cm^{-1}, where it appears to stabilize within 60 min. In comparison, with the plasma/catalyst coupling, the final pH and conductivity in the treated solution are 4.35 and 49.5 μS·cm^{-1} for a treatment of 60 min. The observed changes in the pH can be explained by the production of carboxylic acids in the solution, while the rise of nitrites or nitrates, discussed below, could explain, among other factors, the observed trends of conductivity.

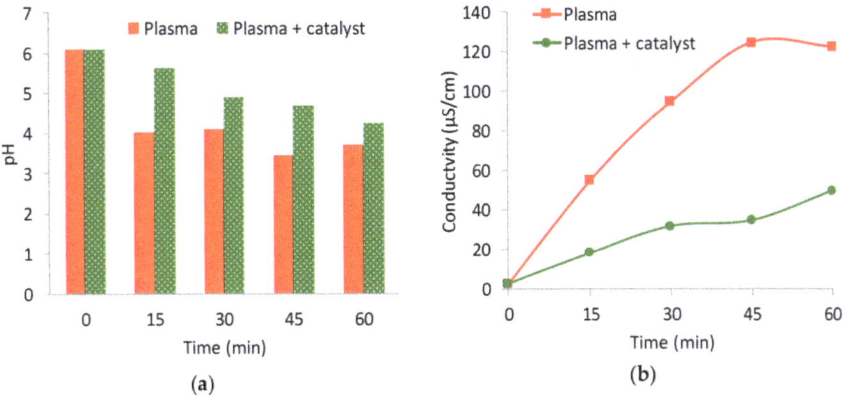

Figure 9. pH (**a**) and conductivity values (**b**) of the solutions treated by NTP/Fenton-like coupling compared to those with NTP alone (U = ±5.9 kV, f = 500 Hz, electrode-liquid-gap = 5 mm, Q_{air} = 100 sccm, [Paracetamol] = 25 mg L^{-1}, V = 40 mL, pH$_0$ = 6.1, σ_0 = 2.61 μS·cm^{-1}).

Nitrites and nitrates are produced during the plasma treatment as displayed in Figure 10. It can be seen that with plasma/catalysis coupling, the concentrations of nitrates and nitrites are lower, or even below the detection threshold which was 4 mg L^{-1} for nitrate (the case of 15 min treatment by NTP/Catalyst), than those with plasma alone.

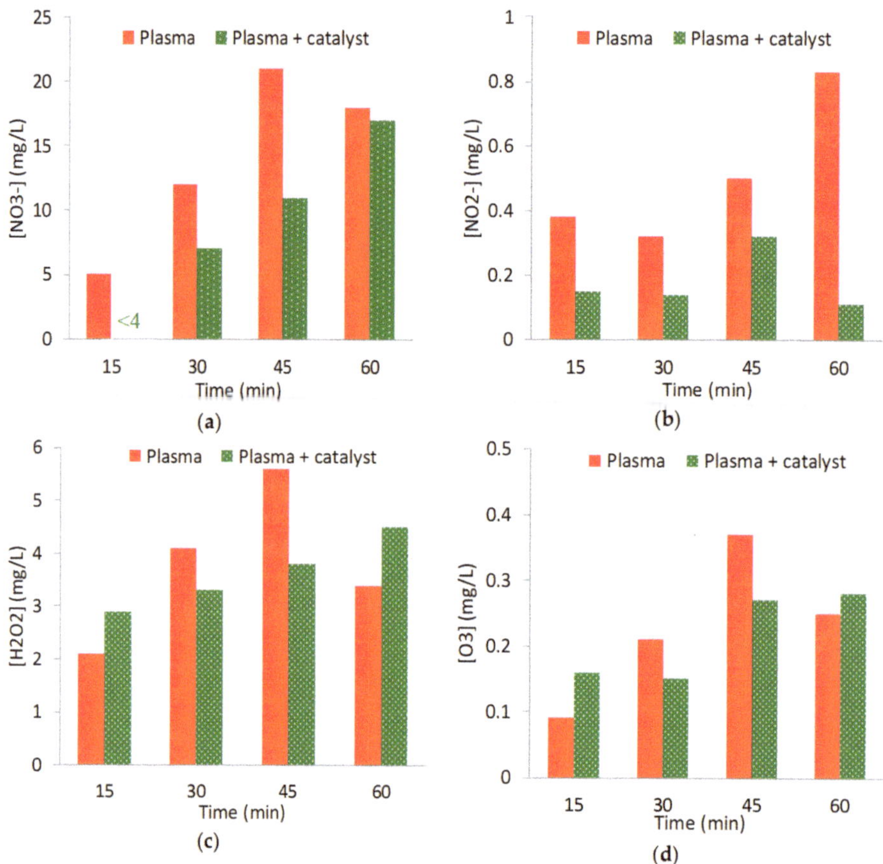

Figure 10. Concentrations of (**a**) Nitrates (4 mg/L is the threshold value), (**b**) Nitrites, (**c**) Hydrogen peroxide, and (**d**) Ozone of the treated samples by NTP/Fenton-like compared with the NTP alone (U = ±5.9 kV, f = 500 Hz, electrode-liquid-gap = 5 mm, Q_{air} = 100 sccm, [Paracetamol] = 25 mg L^{-1}, V = 40 mL).

The generation of nitrites or nitrates can be explained by the following pathways. The excited species in the discharge plasma undergo dissociation, forming NO and NO$_2$ in the gas phase that may be transformed into the inorganic acids HNO$_3$ and HNO$_2$ in liquid (Equations (7)–(12)) [36–38]. Note that the concentrations of these acids depend on many parameters such as the concentration of the excited species formed in the gas phase and their transformation into mineral acids in the liquid phase.

$$e^- + O_2 \rightarrow {}^*O + O + e^- \tag{7}$$

$$e- + N_2 \rightarrow {}^*N + {}^*N + e- \tag{8}$$

$${}^*N + {}^*O \rightarrow NO \tag{9}$$

$${}^*O + NO \rightarrow NO_2 \tag{10}$$

$$NO_2 + H_2O \rightarrow NO_3^- + 2H^+ \tag{11}$$

$$HO^\bullet + NO/NO_2 \rightarrow HNO_2/HNO_3 \tag{12}$$

The presence of NO_3- and NO_2- can be explained by the dissolution of NO_2 in water (Equation (13). Nitrous and nitric acids are highly unstable, and they convert into NO_2- easily (or dissociate into NO_2- or NO_3- (Equation (14)) depending on the pH of the solution [36]:

$$NO_{2(aq)} + NO_{2(aq)} + H_2O_{(l)} \rightarrow NO_2^- + NO_3^- + 2H^+ \tag{13}$$

$$HNO_3/HNO_2 \overset{H_2O}{\Leftrightarrow} H^+ + NO_3^-/NO_2^- \tag{14}$$

H_2O_2 and O_3 were also detected in the treated liquid as shown in Figure 10c,d, respectively. These species are produced by the reactions in the discharge starting from H_2O and O_2 as shown in the equations below [39,40]:

$$^*e^- + H_2O \rightarrow {}^\bullet H + {}^\bullet OH + e^- \tag{15}$$

$$^\bullet OH + {}^\bullet OH \rightarrow H_2O_2 \tag{16}$$

$$H_2O_2 + UV \rightarrow {}^\bullet OH + {}^\bullet OH \tag{17}$$

$$^*O + O_2 + M \rightarrow O_3 + M \tag{18}$$

The oxidative species H_2O_2, $HO^\bullet$, and O_3 are well-known to react with organic molecules. The presence of H_2O_2 in the treated solution confirms the availability of this species to participate in the Fenton-like mechanism in order to promote the generation of $HO^\bullet$ and improve the degradation and mineralization rates of paracetamol and other aromatic by-products. As the Fenton-like effects require H_2O_2 to produce $HO^\bullet$, one may expect a lower concentration of H_2O_2 in the process with the catalyst, but our measurements (Figure 10c) are not in accordance with the mentioned expectation.

According to Equations (19) and (20) [36], H_2O_2 and O_3 in the solution can oxidize nitrite to produce NO_3-, which are considered as blocking reactions for pollutant degradation:

$$NO_2^- + O_3 \rightarrow NO_3^- + O_2 \tag{19}$$

$$NO_2^- + H_2O_2 \rightarrow NO_3^- + H_2O \tag{20}$$

These different sources of hydrogen peroxide production are important for the initiation of the heterogeneous Fenton process in the coupling process with plasma.

2.6. Energy Yield

Typically, in these experiments, the power consumed was 0.31 W with plasma alone and 0.21 W for NTP/catalysis coupling. The Energy Yield (EY) for the plasma alone was in the range of 2.9–4.4 g·(kWh)$^{-1}$, and is shown in Figure 11a as a function of the treatment time, while Figure 11b shows the energy yield as a function of the conversion rate. The EY of the NTP/catalysis coupling system was higher (160 to 200%) than that with plasma alone, indicating that the catalyst significantly improved the EY of the NTP process. Figure 11b showed that the coupling process simultaneously resulted in a better conversion rate and better energy yield. Note that because the calculation used the total dissipated electrical energy, of which only a portion was used for the degradation of paracetamol, so the rate given in this paper is then the mean EY of the process, but not the mean EY of paracetamol degradation.

Even with a much higher input power, Panorel et al. [11] obtained a higher EY (up to 28 g·(kWh)$^{-1}$ with air as working gas) since the treated volume was 40 L, much bigger than the one used in the present work. In the work of Pan and Qiao [14], with a large dissipated power of 500 W and a treated volume of 1.5 L, their EY should be one-tenth of ours. However, carrying out a direct comparison of EY remains difficult since the experimental devices, the treated volumes, and the concentrations of paracetamol are very different.

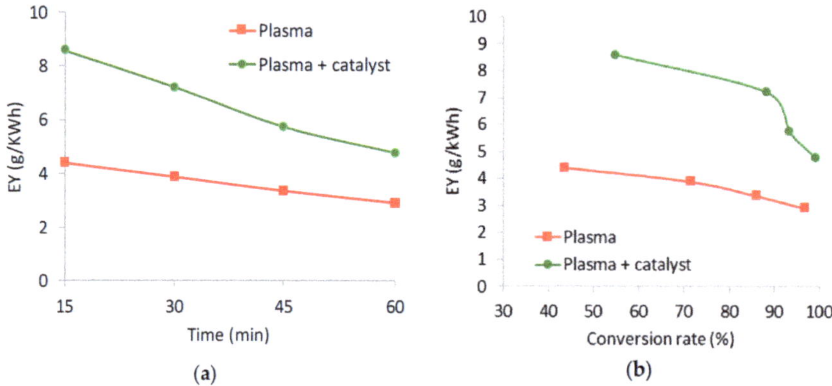

Figure 11. (a) Energy yield (means of 3 measurements) as a function of treatment time and (b) EY vs. conversion rate for NTP alone and NTP/Fenton-like coupling treatments (U = ±5.9 kV, f = 500 Hz, electrode-liquid-gap = 5 mm, Q_{air} = 100 sccm, [Paracetamol] = 25 mg L^{-1}, V = 40 mL).

3. Materials and Methods

3.1. Reagents

Paracetamol (Acetaminophen, BioXtra ≥ 99%), sodium dihydrogen phosphate monohydrate ($NaH_2PO_4 \cdot H_2O$), orthophosphoric acid (H_3PO_4 85%), Aluminum oxide nanopowder (13 μm primary particle size), and iron (III) nitrate nonahydrate ($Fe(NO_3)_3 \cdot 9H_2O$) were supplied by Sigma Aldrich® (France). Acetonitrile, nitric acid (HNO_3 at 70%), and ultrapure water (HPLC gradient) were obtained from Fischer Scientific® (France). All chemicals were used without any additional purification. Commercial standards for HPLC analyses were supplied by Sigma Aldrich® (1,4-benzoquinone Supelco®, hydroquinone ReagentPlus® ≥ 99%, spectrophotometric grade *p*-nitrophenol, maleic acid Reagentplus® ≥ 99%, oxalic acid ≥ 99.0% (RT), acetamide~99% (GC)).

3.2. Catalyst Preparation

Commercial Glass Fiber (GF) tissue, 300 g/m^2, was used as a support for iron oxide deposition via the wet impregnation method. To prepare the supported catalyst, the four steps below were carried out. Note that drying and calcinations were done under ambient air, except for the last calcination which used pure oxygen (in step c).

(a)　GF cleaning: The GF tissue was cut in a circular shape with a diameter of 110 mm. In order to remove residual carbon coming from the fabrication process of the GF, it was calcined at 600 °C for 4 h. The GF was washed repeatedly with distilled water until it no longer released any carbon into the water. The non-release of carbon was checked by measuring the Total Organic Carbon (TOC) of the washing water.

(b)　Washcoating of the GF: The washcoat solution was prepared with 15 g of alumina, 0.7 mL of nitric acid, and 20 mL of ultrapure water. The mixture was kept under vigorous stirring for 12 h at room temperature [41], then GF was dipped in the washcoat solution. The washcoated GF tissue was dried at 100 °C for 12 h then calcined at 400 °C for 4 h. This step creates a bearing layer for the iron precursor.

(c)　Iron impregnation: The alumina-washcoated GF tissue was dipped into an aqueous solution containing 1 g of iron (III) nitrate nonahydrate, and then dried for 16 h at 100 °C. The GF tissue was calcined under an oxygen flow rate of 1 mL·min^{-1} at 400 °C for 4 h.

(d)　Catalyst storage: The final catalyst was washed several times with distilled water then stored in closed glass dishes at 100 °C.

3.3. Catalyst Characterisation Using BET and TGA

The specific surface area of the catalyst was measured using the Brunauer-Emmett-Teller (BET) surface area method. The BET surface area of the materials was determined from nitrogen adsorption at −196 °C obtained with an ASAP 2020 Micromeritics apparatus. Beforehand, the samples were degassed at 250 °C for at least 24 h under a residual vacuum of at least 6 μm Hg (0.8 Pa).

Table 1 shows the specific BET surface area (S_{BET}) of Fe/GF before and after 90 min of use in the plasma/catalysis coupling process. It can be seen that there is a very slight increase in the surface area of the catalyst after use.

Table 1. $S_{(BET)}$ of Fe/GF before and after 90 min of use in the plasma/catalysis coupling process.

	Virgin Glass Fiber	Catalyst (Before Use)	Catalyst (After Use)
$S_{(BET)}$ (m^2/g)	0	10	14

Thermogravimetric analyses were performed using a Netzsch STA 449 JUPITER F5. A mass of approximately 0.040 g of sample was placed in the thermobalance under a 160 mL min^{-1} argon flow. Systematically, a blank experiment was carried out, under the same conditions, in order to determine the correction required for the base-line. The procedure consisted of a temperature increase at a 20 °C min^{-1} heating ramp from ambient temperature to 900 °C (to study a larger temperature domain), followed by cooling to room temperature at a cooling rate of 10 °C min^{-1}. The TGA analyses (not shown here) indicated that there was no difference in the thermogram of each material studied with a mass loss of approximately 1%.

3.4. Experimental Setup

The Non Thermal Plasma (NTP) reactor shown in Figure 12 was already described in a previous paper [13]. Briefly, a Dielectric Barrier Discharge (DBD) reactor with multiple needle-to-plate configurations was placed inside a cylindrical recipient made of Polyvinyl Chloride (PVC). High-Voltage (HV) was applied to the 12 stainless steel needle electrodes with 0.4 mm of inner diameter. The working gas (air) was introduced through the hollow electrodes at a flow rate of 100 sccm. The grounded electrode was a copper thin film deposited on a dielectric epoxy plate. The volume of the solution of paracetamol to be treated was 40 mL prepared in ultrapure water at a concentration of 25 mg L^{-1} maintaining a distance between the electrodes and the surface of the liquid of 5 mm.

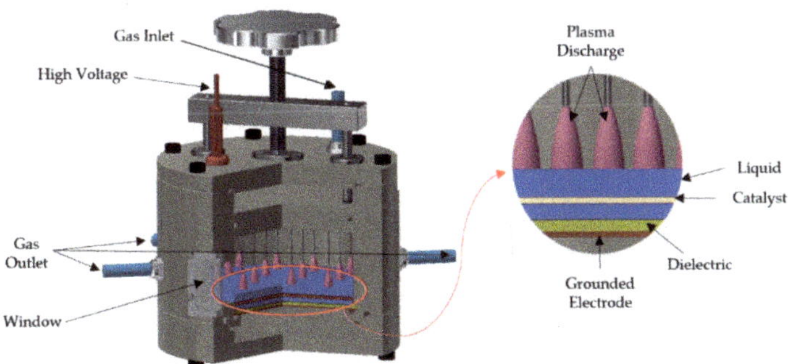

Figure 12. Scheme of catalyst coupled non-thermal plasma reactor.

For NTP/Catalysis coupling, the catalyst was dipped into the liquid above the dielectric, as illustrated in the inset in Figure 12. The discharges were generated above the liquid by a HV amplifier (Trek® 20/20C) with a frequency of 500 Hz driven by a function generator (TTI® TG4001). Electrical measurements were performed using high-voltage (PMK-14KVAC PHV4-3272) and current

(P6015A Tektronix®) probes connected to an oscilloscope (DPO 3054 Tektronic®). A capacitor of 0.5 nF was inserted in the discharge circuit in order to determine the dissipated power using the Lissajous method.

3.5. Analytical Methods & Instrumentation

A High Performance Liquid Chromatography system (HPLC, NEXERA i-Series LC2040C 3D, Shimadzu Corporation, France) with an SPD-40V UV-Vis detector was used to detect paracetamol and its intermediates. The analyte separation was carried out on a 150 × 3.0 mm reverse-phase Shim-pack GIST-HP C18 (Shimadzu®) column with a spherical particle (Octadecyl groups) size of 3 μm diameter. The column temperature was set at 40 °C. The mobile phase (5:95 v/v) was a mixture of Acetonitrile (ACN): Ultrapure water (20 mM $NaH_2PO_4 \cdot H_2O$) at pH 2.8. The flow rate was 0.6 mL·min⁻¹ and a programmed gradient system was used (5% of ACN for 0.5 min. then ACN increased up to 70% until 10 min and kept stabilized until 13 min; finally, a decrease of ACN down to 5% at 13.5 min). To detect paracetamol molecules, the UV-detection wavelength was 246 nm with a running time of 20 min per sample to determine the retention time and peak area of paracetamol. The concentration of paracetamol residue was calculated using a pre-determined calibration curve.

The paracetamol conversion rate τ (%) for a given treatment time was calculated from the following (Equation (21)):

$$\tau = \left(1 - \frac{C_t}{C_0}\right) \times 100 \tag{21}$$

where C_0 and C_t are the concentrations of paracetamol in mg L⁻¹ before and after treatment duration t.

The mineralization of paracetamol was characterized by Total Organic Carbon (TOC) removal, which was determined by a TOC-L$_{CSH/CSN}$ (Shimadzu®) analyzer based on the high-temperature combustion oxidation method. The removal ratio of TOC (τ_{TOC}), indicating the mineralization rate, was calculated by (Equation (22)):

$$\tau_{TOC} = \left(1 - \frac{[TOC]t}{[TOC]_0}\right) \times 100 \tag{22}$$

where [TOC] is the initial TOC of the paracetamol solution in (mg L⁻¹), [TOC]t is the TOC value at treatment time t (min).

The Energy Yield of the process was evaluated from the following expression (EY):

$$EY\left(\frac{g}{kWh}\right) = \frac{(C_0 - C_t) \times V}{\Delta t.P} \tag{23}$$

where C_0 and C_t are the concentrations of the pollutant before and after treatment, Δt (in h) is the treatment time; V is the volume of the treated solution (L) and P is the power dissipated in the discharge (in kW). EY corresponds to the mass of paracetamol removed per kWh.

The concentrations of nitrites, nitrates, hydrogen peroxide, and ozone in aqueous phase were measured with a photometer MultiDirect (Lovibond®). The values of pH and conductivity of the solutions before and after treatment were measured with an Accumet AB200 (Fisher Scientific®).

Author Contributions: Conceptualization, N.K., O.A., H.R. and D.H.; Funding acquisition, O.A. and D.H.; Investigation, N.K., O.A., H.R. and B.C.; Project administration, O.A., and D.H.; Resources, O.A., H.R. and D.H.; Supervision, O.A., H.R. and D.H.; Validation, N.K., O.A., H.R., B.C. and D.H.; Visualization, N.K., O.A. and D.H.; Writing–original draft, N.K., O.A., H.R., B.C. and D.H. All authors have read and agreed to the published version of the manuscript.

Funding: This research was partial supported by the PIVOTS project by the Région Centre–Val de Loire (ARD 2020 program and CPER 2015 -2020) and the French Ministry of Higher Education and Research (CPER 2015–2020 and public service subsidy to CNRS and Université d'Orléans). N.K. thanks the Région Centre–Val de Loire for her PhD scholarship.

Acknowledgments: The authors are very grateful to Caroline Norsic for discussion on catalyst preparation.

Conflicts of Interest: The authors declare no conflict of interest.

References

1. Hillis, D.G.; Lissemore, L.; Sibley, P.K.; Solomon, K.R. Effects of Monensin on Zooplankton Communities in Aquatic Microcosms. *Environ. Sci. Technol.* **2007**, *41*, 6620–6626. [CrossRef]
2. Kim, Y.; Choi, K.; Jung, J.; Park, S.; Kim, P.-G.; Park, J. Aquatic toxicity of acetaminophen, carbamazepine, cimetidine, diltiazem and six major sulfonamides, and their potential ecological risks in Korea. *Environ. Int.* **2007**, *33*, 370–375. [CrossRef]
3. Talib, A.; Randhir, T.O. Managing Emerging Contaminants: Status, Impacts, and Watershed-Wide Strategies. *Expo. Health* **2016**, *8*, 143–158. [CrossRef]
4. Saeid, S.; Tolvanen, P.; Kumar, N.; Eränen, K.; Peltonen, J.; Peurla, M.; Mikkola, J.-P.; Franz, A.; Salmi, T. Advanced oxidation process for the removal of ibuprofen from aqueous solution: A non-catalytic and catalytic ozonation study in a semi-batch reactor. *Appl. Catal. B Environ.* **2018**, *230*, 77–90. [CrossRef]
5. Van Doorslaer, X.; Dewulf, D.; De Maerschalk, J.; Van Langenhove, H.; Demeestere, K. Heterogeneous photocatalysis of moxifloxacin in hospital effluent: Effect of selected matrix constituents. *Chem. Eng. J.* **2015**, *261*, 9–16. [CrossRef]
6. Tokumura, M.; Sugawara, A.; Raknuzzaman, M.; Habibullah-Al-Mamun, M.; Masunaga, S. Comprehensive study on effects of water matrices on removal of pharmaceuticals by three different kinds of advanced oxidation processes. *Chemosphere* **2016**, *159*, 317–325. [CrossRef]
7. Gomes, J.; Costa, R.; Quinta-Ferreira, R.M.; Martins, R.C. Application of ozonation for pharmaceuticals and personal care products removal from water. *Sci. Total Environ.* **2017**, *586*, 265–283. [CrossRef]
8. Bokare, A.D.; Choi, W. Review of Iron-Free Fenton-Like Systems for Activating H_2O_2 in Advanced Oxidation Processes. *J. Hazard. Mater.* **2014**, *275*, 121–135. [CrossRef]
9. Klavarioti, M.; Mantzavinos, D.; Kassinos, D. Removal of residual pharmaceuticals from aqueous systems by advanced oxidation processes. *Environ. Int.* **2009**, *35*, 402–417. [CrossRef]
10. Ajo, P.; Preis, S.; Vornamo, T.; Mänttäri, M.; Kallioinen, M.; Louhi-Kultanen, M. Hospital wastewater treatment with pilot-scale pulsed corona discharge for removal of pharmaceutical residues. *J. Environ. Chem. Eng.* **2018**, *6*, 1569–1577. [CrossRef]
11. Panorel, I.; Preis, S.; Kornev, I.; Hatakka, H.; Louhi-Kultanen, M. Oxidation of aqueous pharmaceuticals by pulsed corona discharge. *J. Environ. Technol.* **2013**, *34*, 923–930. [CrossRef]
12. Magureanu, M.; Mandache, B.N.; Parvulescu, V.I. Degradation of pharmaceutical compounds in water by non-thermal plasma treatment. *Water Res.* **2015**, *81*, 124–136.
13. Baloul, Y.; Aubry, O.; Rabat, H.; Colas, C.; Maunit, B.; Hong, H. Paracetamol degradation in aqueous solution by non-thermal plasma. *Eur. Phys. J. Appl. Phys.* **2017**, *79*, 1–7. [CrossRef]
14. Pan, X.; Qiao, X. Influences of nitrite on paracetamol degradation in dielectric barrier discharge reactor. *Ecotoxicol. Environ. Saf.* **2019**, *180*, 610–615. [CrossRef]
15. Iervolino, G.; Vaiano, V.; Palma, V. Enhanced removal of water pollutants by dielectric barrier discharge nonthermal plasma reactor. *Sep. Purif. Technol.* **2019**, *215*, 155–162. [CrossRef]
16. Xu, L.; Wang, J. Fenton-like degradation of 2,4-dichlorophenol using Fe_3O_4 magnetic nanoparticles. *Appl. Catal. B Environ.* **2012**, *123*, 117–126. [CrossRef]
17. Guélou, E.; Barrault, J.; Fournier, J.; Tatibouët, J.-M. Active iron species in the catalytic wet peroxide oxidation of phenol over pillared clays containing iron. *Appl. Catal. B Environ.* **2003**, *44*, 1–8. [CrossRef]
18. Legube, B.; Leitner, N.K.V. Catalytic ozonation: A promising advanced oxidation technology for water treatment. *Catal. Today* **1999**, *53*, 61–72. [CrossRef]
19. Kasprzyk-Hordern, B.; Ziółek, M.; Nawrocki, J. Catalytic ozonation and methods of enhancing molecular ozone reactions in water treatment. *Appl. Catal. B Environ.* **2003**, *46*, 639–669. [CrossRef]
20. Du, Y.; Zhou, M.; Lei, L. The role of oxygen in the degradation of p-chlorophenol by Fenton system. *J. Hazard. Mater.* **2007**, *139*, 108–115. [CrossRef]
21. Xu, Z.; Xue, X.; Hu, S.; Li, Y.; Shen, J.; Lan, Y.; Zhou, R.; Yang, F.; Cheng, C. Degradation effect and mechanism of gas-liquid phase dielectric barrier discharge on norfloxacin combined with H_2O_2 or Fe^{2+}. *Sep. Purif. Technol.* **2020**, *230*, 1–11. [CrossRef]

22. Li, X.; Chen, W.; Ma, L.; Wang, H.; Fan, J. Industrial wastewater advanced treatment via catalytic ozonation with an Fe-based catalyst. *Chemosphere* **2018**, *195*, 336–343. [CrossRef]

23. Reddy, P.M.K.; Dayamani, A.; Mahammadunnisa, S.; Subrahmanyam, C. Mineralization of Phenol in Water by Catalytic Non-Thermal Plasma Reactor—An Eco-Friendly Approach for Wastewater Treatment. *Plasma Process. Polym.* **2013**, *10*, 1010–1017. [CrossRef]

24. Kušić, H.; Koprivanac, N.; Locke, B.R. Decomposition of phenol by hybrid gas/liquid electrical discharge reactors with zeolite catalysts. *J. Hazard. Mater.* **2005**, *125*, 190–200. [CrossRef]

25. Baloul, Y.; Rabat, H.; Hong, D.; Chuon, S.; Aubry, O. Preliminary Study of a Non-thermal Plasma for the Degradation of the Paracetamol Residue in Water. *Int. J. Plasma Environ. Sci. Technol.* **2016**, *10*, 102–107.

26. Nie, Y.; Zhang, L.; Li, Y.-Y.; Hu, C. Enhanced Fenton-like degradation of refractory organic compounds by surface complex formation of LaFeO$_3$ and H$_2$O$_2$. *J. Hazard. Mater.* **2015**, *294*, 195–200. [CrossRef]

27. Yi, Q.; Ji, J.; Shen, B.; Dong, C.; Liu, J.; Zhang, J.; Xing, M. Singlet Oxygen Triggered by Superoxide Radicals in a Molybdenum Cocatalytic Fenton Reaction with Enhanced REDOX Activity in the Environment. *Environ. Sci. Technol.* **2019**, *53*, 9725–9733. [CrossRef]

28. Walling, C. Fenton's Reagent Revisited. *Acc. Chem. Res.* **1975**, *8*, 125–131. [CrossRef]

29. Wang, S. A Comparative study of Fenton and Fenton-like reaction kinetics in decolourisation of wastewater. *Dye. Pigment.* **2008**, *76*, 714–720. [CrossRef]

30. Brillas, E.; Sirés, I.; Cabot, P.L.; Centellas, F.; Rodríguez, R.M.; Garrido, J.A. Mineralization of paracetamol in aqueous medium by anodic oxidation with a boron-doped diamond electrode. *Chemosphere* **2005**, *58*, 399–406. [CrossRef]

31. Slamani, S.; Abdelmalek, F.; Ghezzar, M.R.; Addou, A. Initiation of Fenton process by plasma gliding arc discharge for the degradation of paracetamol in water. *J. Photochem. Photobiol. A Chem.* **2018**, *359*, 1–10. [CrossRef]

32. Yan, J.; Du, C.; Li, X.; Sun, X.; Ni, M.; Cen, K.; Cheron, B. Plasma chemical degradation of phenol in solution by gas-liquid gliding arc discharge. *Plasma Sources Sci. Technol.* **2005**, *14*, 637–644. [CrossRef]

33. Garrido, J.A.; Brillas, E.; Cabot, P.L.; Centellas, F.; Arias, C.; Rodríguez, R.M. Mineralization of drugs in aqueous medium by advanced oxidation processes. *Port. Electrochim. Acta* **2007**, *25*, 19–41. [CrossRef]

34. Arredondo Valdez, H.C.; Jiménez, G.G.; Granados, S.G.; De León, C.P. Degradation of paracetamol by advance oxidation processes using modified reticulated vitreous carbon electrodes with TiO$_2$ and CuO/TiO$_2$/Al$_2$O$_3$. *Chemosphere* **2012**, *89*, 1195–1201. [CrossRef] [PubMed]

35. Moctezuma, E.; Leyva, E.; Aguilar, C.A.; Luna, R.A.; Montalvo, C. Photocatalytic degradation of paracetamol: Intermediates and total reaction mechanism. *J. Hazard. Mater.* **2012**, *243*, 130–138. [CrossRef]

36. Lukeš, P.; Dolezalova, E.; Sisrova, I.; Clupek, M. Aqueous-phase chemistry and bactericidal effects from an air discharge plasma in contact with water: Evidence for the formation of peroxynitrite through a pseudo-second-order post-discharge reaction of H$_2$O$_2$ and HNO$_2$. *Plasma Sources Sci. Technol.* **2014**, *23*, 1–15. [CrossRef]

37. Burlica, R.; Kirkpatrick, M.J.; Locke, B.R. Formation of reactive species in gliding arc discharges with liquid water. *J. Electrost.* **2006**, *64*, 35–43. [CrossRef]

38. Chen, G.; Zhou, M.; Chen, S.; Chen, W. The different effects of oxygen and air DBD plasma byproducts on the degradation of methyl violet 5BN. *Hazard. Mater.* **2009**, *172*, 786–791. [CrossRef]

39. Huang, F.; Chen, L.; Wang, H.; Yan, Z. Analysis of the degradation mechanism of methylene blue by atmospheric pressure dielectric barrier discharge plasma. *Chem. Eng. J.* **2010**, *162*, 250–256. [CrossRef]

40. Reddy, P.M.K.; Raju, B.R.; Karuppiah, J.; Reddy, E.L.; Subrahmanyam, C.H. Degradation and mineralization of methylene blue by dielectric barrier discharge non-thermal plasma reactor. *Chem. Eng. J.* **2013**, *217*, 41–44. [CrossRef]

41. Villegas, L.; Masset, F.; Guilhaume, N. Wet impregnation of alumina-washcoated monoliths: Effect of the drying procedure on Ni distribution and on autothermal reforming activity. *Appl. Catal. A Gen.* **2007**, *320*, 43–55. [CrossRef]

MDPI
St. Alban-Anlage 66
4052 Basel
Switzerland
Tel. +41 61 683 77 34
Fax +41 61 302 89 18
www.mdpi.com

Catalysts Editorial Office
E-mail: catalysts@mdpi.com
www.mdpi.com/journal/catalysts